P9-CRY-116

FLOWERING PLANTS IN THE LANDSCAPE

FLOWERING PLANTS IN THE LANDSCAPE

Mildred E. Mathias, Editor

Foreword by Sir George Taylor

UNIVERSITY OF CALIFORNIA PRESS

Berkeley Los Angeles London

University of California Press
Berkeley and Los Angeles, California

University of California Press, Ltd.
London, England

Library of Congress Cataloging in Publication Data
Main entry under title:

Flowering plants in the landscape.

Bibliography: p. 237
Includes index.
1. Flowering woody plants—Tropics—Pictorial works. 2. Ground cover plants—Tropics—Pictorial works. 3. Tropical plants—Pictorial works. 4. Flowers—Tropics—Pictorial works. 5. Plants, Ornamental—California—Pictorial works. 6. Color in gardening. 7. Landscape gardening—Tropics. 8. Wild flower gardening—California. I. Mathias, Mildred Esther, 1906-
SB435.6.T76F58 635.9'6 81-16310
ISBN 0-520-04350-2 AACR2

Printed in the United States of America
1 2 3 4 5 6 7 8 9

CONTENTS

FOREWORD

It is a valued privilege to introduce a new enlarged edition of this highly esteemed work. The original version was widely acclaimed by landscape architects, ordinary gardeners and, indeed, by all who appreciate splendid illustrations to accompany informed text. The earlier volume has greatly fostered an intelligent interest in deciding the best choice of plants for color from ground cover to sizable trees for gardens in tropical and subtropical regions. Those who endure the seasonal variation and often privation of temperate zones envy the almost bewildering range of superb species that provide radiance throughout the year in gardens in these more favored areas.

The selection of elite plants is widely culled from the warmer parts of the world, and the striking richness of the California native flora is well represented. But the appeal of this volume will not be confined to growers in California. It is emphatically such a useful guide to the colorful ornamental plants seen in gardens and in the wild and in the tropics and subtropics that it should become a handbook for travelers. Apart from the simple descriptions identification is greatly assisted by the beautiful plates. This book is a treasure in its genre and cannot fail to bring aesthetic and intellectual enjoyment to its user.

Sir George Taylor

Former Director
Royal Botanic Gardens, Kew

PREFACE

This book is a complete revision enlarged to include additional species and many of the more common and showy tropical plants. Many new color photographs have also been added. In 1964 Los Angeles Beautiful, California Arboretum Foundation, and the Southern California Horticultural Institute published a booklet on flowering trees. This was followed by booklets on flowering shrubs, vines, and ground covers. In 1967 Los Angeles Beautiful produced a booklet on *Erythrina,* the official tree of the city of Los Angeles. The Theodore Payne Foundation for Wild Flowers and Native Plants joined the other sponsors in 1971 to publish a booklet on Colorful California Native Plants. These six booklets were revised and expanded and in 1973 the sponsors produced Color for the Landscape: Flowering Plants for Subtropical Climates. In 1976 two new sponsors, Descanso Gardens Guild and South Coast Botanic Garden Foundation, joined in the publication of a second updated printing. The efforts of these several sponsors and of their representatives on the Flowering Tree Committee of Los Angeles Beautiful have led to the present expanded volume.

ACKNOWLEDGMENTS

Many individuals have contributed through the years to the publications culminating in the present volume. These include the members of the Flowering Tree Committee of Los Angeles Beautiful who have continually promoted the use of flowering plants in the landscape and the chairman, Dr. Samuel Ayres, Jr., author of the original text on flowering shrubs. Their support has been invaluable. Others who have contributed include:

Dr. Robert E. Atkinson, author, botanical consultant
Virginia M. Baldwin, Los Angeles Beautiful
Fred Boutin, botanist
Betty Thomas Carriel
Philip E. Chandler, Chandler & Lang, Landscape Planning, Inc., author of the original texts on flowering trees and vines
Francis Ching, director, Los Angeles County Department of Arboreta and Botanic Gardens
Mrs. Ralph D. Cornell
Henry Davis, publisher of Flowering Booklet series
Merritt S. Dunlap, former president, Theodore Payne Foundation, Inc.
Dr. Leonid Enari, taxonomist, Los Angeles State and County Arboretum
Morgan Evans, A.I.L.A.
Walter C. Hughes, Jr.
Myron Kimnach, curator, Huntington Botanical Gardens, San Marino
Dr. Lee W. Lenz, director, Rancho Santa Ana Botanic Garden, Claremont
Elisabeth Marshall
Dr. Elizabeth McClintock
Dr. Katherine K. Muller, director emeritus, Santa Barbara Botanic Garden
James C. Perry, Perry's Plants, Inc.
Edward L. Peterson, Theodore Payne Foundation, Inc., author of original text on California native plants
George H. Spalding
Dr. William S. Stewart
Dr. Vernon T. Stoutemyer, chairman emeritus, Department of Agricultural Sciences, University of California, Los Angeles, author of the original text on ground covers
David S. Verity, senior museum scientist, University of California, Los Angeles
Donald P. Woolley, former superintendent, South Coast Botanic Garden

Photographs by Ralph D. Cornell, FASLA except for the following:

William Aplin, pp. 10 inset, 22, 31, 33, 34, 88, 89, 90, 92, 94, 107, 130, 131, 147, 158, 165, 166, 167, 168;
Dr. Robert E. Atkinson, pp. 68, 114, 126, 136;
Dr. Samuel Ayres, Jr., pp. 21, 27, 93;
Kenneth Budlong, p. 182;
A. W. and M. V. Hood, pp. 194 upper right, 195, 204, 208;
Mildred E. Mathias, pp. 130, 200, 201, 206;
Nuccio's Nursery, p. 64;
Virginia J. Robertson, p. 66;
David S. Verity, pp. 129, 161, 163, 198, 203, 205, 206;
Sue Verity, p. 172;
Don Walker, p. 164;
Laurel Woodley, pp. 191, 194 upper left and bottom, 204.

Ralph Dalton Cornell, F.A.S.L.A., 1890-1972

To the memory of *Ralph Dalton Cornell,* dean of landscape architects, inspired plantsman and photographer—whose creative talent, intellect and good taste brought beauty and dignity to everything he touched—this book is dedicated with gratitude.

BIOGRAPHICAL NOTE

Ralph Dalton Cornell was born in Holdrege, Nebraska, on January 11, 1890. His father was considered a successful lumberman and also owned several outlying ranches. While the family lived in town, young Ralph—always on horseback—roamed the expanses of the surrounding countryside and early fell in love with the great outdoors.

He was just sixteen when the Cornells visited the resort town of Long Beach in California. They were entranced with the beauties of the land and, as many others before them, decided to make the "big move" and settled in Long Beach in 1908.

One of the primary attractions for lumberman Cornell was the promised wealth of the flourishing eucalyptus industry that had captured the imaginations of many Californians at the time. Unfortunately, the promoters knew far too little about the growing of the Australian native and the processing of its rather capricious wood. Cornell Senior joined the hordes who lost their life savings owing to the sudden collapse of the eucalyptus boom.

But the break with Nebraska had been made and Ralph and his two sisters entered Long Beach High School. Instead of matriculating at the University of Nebraska as he had hoped, Ralph eventually entered Pomona College in Claremont in 1909. It was necessary for him to work his way through college. He waited on table at the old Claremont Inn. In 1910 he acquired 1,000 avocado seeds, propagated them successfully and sold the exotic plants at a good profit. He surveyed the agricultural development of the Coachella and Imperial valleys for an irrigation association headed by George Wharton James in July of 1911. He spent another hot summer vacation helping the Popenoe brothers plant 13,000 date offshoots they had imported from Arabia to begin their nursery a few miles south of Thermal.

These two latter experiences, while horrendous because of the extreme summmer heat, served to introduce him to the mysterious and wondrous plant life of the desert, which he never forgot and which never ceased to allure him.

Cornell's prime interests were botany and horticulture but his mentor while at Pomona College, Professor Charles Fuller Baker, urged him to consider the fairly new profession of landscape architecture. It was Professor Baker who was responsible for the publication of the *Pomona College Journal of Economic Botany* and it was in this journal that Cornell's first published article entitled "Plans and Plants for Small Places" appeared in 1911. The inclusion of "Plans" surely indicated that he was strongly considering Professor Baker's recommendation.

After his graduation from Pomona College, *summa cum laude* and Phi Beta Kappa, Cornell entered the School of Design at Harvard University. His three years in this eminent institution provided some of his happiest memories.

The position he accepted in Canada upon his graduation from Harvard was terminated when he volunteered for overseas army duty in World War I. He detested violence all the days of his life but he loved his country and he was imbued with a great sense of patriotism.

He returned to Los Angeles in 1919 where he opened an office and his first commission was that of supervising landscape architect for Pomona College. This led to a request from the University of Hawaii for a master plan for their campus. And, eventually, Robert Gordon Sproul, president of the University of California, invited Cornell to supervise the landscape of the new Westwood campus. This association contined for almost forty years, terminating with Cornell's death.

His partnership with Wilbur D. Cook, the well-known landscape architect who designed the master plan for the city of Beverly Hills in 1906, was responsible for Cornell's long affiliation and extensive contributions to the beauty of that community.

These commissions were the beginnings of an illustrious career. Cornell's outstanding design ability, his distinctive taste and style, his special vision and sensitivity to the particular needs of each undertaking soon won for him an enviable reputation.

The scope of his work was astonishing. He did public parks and colleges; he did cemeteries and subdivisions; he did civic centers and malls. He did master plans for Elysian and Griffith parks in Los Angeles and for the Torrey Pines Preserve in La Jolla. He did Hilton Hotels in Egypt and Iraq. He planned the town of Valencia in southern California. He did Fremont Place in Los Angeles in 1912 and almost fifty years later the prestigious Music Center of Los Angeles County.

His legacy was an infinite variety of spectacular and beautiful designs.

Cornell was fortunate that his chosen profession allowed him to indulge in his favorite avocations of botany and photography. His lifelong love of plants and his expertise in horticulture were great assets and impelled him to use both exotics and natives in his landscape design. This has left us the richer for his choice of plantings are botanical gardens of rare treasures.

The sky, the sun, the moon, the stars, the earth itself, and always the plants that grow in the earth—these provided zest and meaning for his entire life.

Ralph Cornell was a humble, unassuming man who gave of himself and his special talents with unstinting generosity for the continued pleasure and gratification of all.

Vera B. Cornell

BIOGRAPHICAL NOTE

Largely owing to the efforts of one man, California wild flowers are found in gardens all over the world. This man was Theodore Payne in whose honor a California native plant foundation was established in 1960. The Theodore Payne Foundation operates a nursery, art gallery, and library on twenty-two acres at 10459 Tuxford Street, Sun Valley, California.

When Theodore Payne was a young lad hunting wild flowers in the hills around Northampshire, England, an event occurred at the Royal Botanic Gardens at Kew which was to shape his entire life. An exhibit of plants collected by Archibald Menzies in far-off California drew large crowds and captured the hearts of avid English gardeners. When he was twenty-one Payne headed for California where he found the people largely unaware of the bountiful floral array nature had bestowed. He immediately set about collecting seeds and heralding the native flora.

Theodore Payne brought some 430 native species into cultivation and through his efforts seed was sent throughout the world. Many native plantings, including Exposition Park in Los Angeles, were done by Payne. He was much interested in Ralph Cornell's development of Torrey Pines Park and the campus of Pomona College. Payne selected the original site of Rancho Santa Ana Botanic Garden and inspired the establishment of the Santa Barbara Botanic Garden. Many honors came to Theodore Payne before his death in 1963. The world has benefited from his contributions to gardens.

Theodore Payne

EDITOR'S INTRODUCTION
FLOWERING PLANTS IN THE LANDSCAPE

This is a book for the adventurous gardener in tropical or subtropical regions and a guide for the flower lover traveling in the tropics and subtropics. Gardening in these regions is a continuing challenge with a wealth of plant materials available for cultivation. Some of the most beautiful flowering plants from the far corners of the world provide a floral display possible only under glass in cooler climates.

The purpose of this book is to present some of the more outstanding colorful plants from around the world, such as *Jacaranda* from South America; many species of *Erythrina* or coral trees; *Acacia* and flowering *Eucalyptus* from Australia; *Protea* and *Leucospermum* from South Africa; and native California plants such as *Ceanothus, Fremontodendron,* and *Romneya.* It is an exciting list of possibilities for enhancing the landscape with the beauty found only in the natural color of flowers.

The book is illustrated with color photographs for the most part by the late Ralph D. Cornell. Each plant illustrated is identified by botanical name, common name, plant family, and area of origin. The minimum temperature that will result in plant damage is given as well as the usual season of bloom or, in the case of plants with colorful fruit, the season of maximum color. The descriptions include size and shape, whether evergreen or deciduous, flower color, microclimate to which best adapted, and special cultural requirements. Where plants are tender or marginal for the subtropics, or so difficult that they are recommended only for the adventurous or advanced gardener, that is indicated also.

Not every plant will grow well and bloom in every garden. Variations in soils and microclimates must be considered in choosing a specific plant for a site. Some plants will flower only in areas with continued heat while others require the moderate cooler coastal or higher elevation climates. Some are best planted against a wall to provide reflected heat. Some need winter chilling to produce flower buds; others need to be kept dry for a dormant period. Most plants do best with good drainage. If your garden has special problems it is best to consult a landscape architect, designer, or your local nurseryman.

For convenience plants are grouped in chapters by plant form or use—as trees, shrubs, vines, and ground covers. However, plants may fit in more than one category. Some trees make fine shrubs; many shrubs with proper pruning make small trees. Some genera because of their variety of form have been placed in subchapters between trees and shrubs. Some shrubs are sprawlers and make excellent vine substitutes while others are best suited for ground covers. The California native plants often require special garden treatment and they are discussed together regardless of form. The reader will find it worthwhile to explore the entire book including the lists of additional plants of merit in the appendixes. The plants in this book offer new color excitement.

Many of the plants best known because they are grown widely in several climatic zones, such as roses and popular annuals and perennials, have been omitted to provide space for discussion of those especially well adapted to subtropical regions. Many species of various genera such as *Eucalyptus, Acacia, Escallonia, Arctostaphylos, Buddleia,* and others, also have been omitted. For descriptions of these and for more cultural details, cold tolerances, and other horticultural information a number of regional books are available.

LEARN THE PROPER NAMES

An important factor in getting the specific plant desired for your landscaping objective is to LEARN THE PROPER NAME.

If the landscape design calls for a thirty-foot acacia to provide a giant bouquet of yellow flowers in late winter you must specify *Acacia baileyana.* If you ask your nurseryman only for an acacia you may get the equally beautiful *Acacia glandulicarpa,* a spring-blooming compact shrub, or any one of the other acacias in the nursery. More disappointing would be to find out at long last that the mimosa tree you bought thinking it to be the yellow-flowered acacia known as mimosa in the part of the world you came from was the pink-flowered *Albizia julibrissin,* called mimosa by your nurseryman.

Knowing the botanical name can prevent such mistakes and disappointments. Botanical names are an international language and refer to the same plant wherever they are used, while common names vary from one part of the world to another. The use of mimosa and acacia is a good example. The wild lilac of California is *Ceanothus* and not the common lilac of gardens, *Syringa,* in a very different plant family. Jasmine is a common name for the genus *Jasminum* in the olive family but star jasmine is a common name for *Trachelospermum jasminoides* in the dogbane family.

Botanical names are not difficult. They normally consist of two parts, the first is the genus and the second the species name. For example, *Acacia* is the genus name and *baileyana* and *glandulicarpa* are names for two different species of *Acacia* illustrated in this book. Both names must be used and the plants are referred to as *Acacia baileyana* and *Acacia glandulicarpa.* In text where the meaning is obvious the genus name may be abbreviated and the botanical name may be written *A. baileyana* or *A. glandulicarpa.* On occasions the name consists of three parts as in *Eucalyptus leucoxylon macrocarpa* or *Camellia sasanqua* 'Yuletide.' In the case of the *Eucalyptus* the species name is *leucoxylon* and *macrocarpa* is a botanical variety of that species. In the case of the *Camellia* the third name 'Yuletide' is that of a cultivated variety, a cultivar. Botanical names are italicized in the text while cultivar names are not italicized but are capitalized and set in single quotation marks. In a name such as *Clematis* X *lawsoniana,* the X indicates that the plant is of hybrid origin. Some cultivars are selections from hybrids that have not been named so that no species name is used as, for example, *Dimorphotheca* 'Buttersweet." Many a genus name may have become familiar through its use as a common name such as hibiscus, fuchsia, acacia, and eucalyptus. When it is used as a common name, it is not capitalized and not italicized.

In the classification of plants similar genera are grouped into a family. Since plant families are often easily recognized by the gardener, even though the genus may not be known, the family name has been included with the descriptions. For example the genera *Acacia, Calliandra, Bauhinia, Cassia,* and many others in this book are members of the Leguminosae or Fabaceae, a large family of plants that also includes such common garden vegetables as peas and beans. The entire family is characterized by its fruit which generally resembles a bean or pea pod.

Most nurseries will have the plants labeled with both the botanical name and local common names. Popular or common names of plants used throughout this book denote those most commonly heard. It must be reemphasized that common names can be misleading. For example the tree called "coral tree" in California is known as "flame tree" in some countries. And there are completely unrelated trees with red flowers also known as "flame tree." Sometimes the common name refers to many species in a genus. Knowing the correct botanical name will make it possible for you to get the plant you want.

I
FLOWERING TREES

Erythrina caffra (see p. 54)

Flowering trees in a riot of color are typical of the tropical regions of the world. They adorn not only the native forests but city streets and home gardens. Many do not realize that, with proper selection from a wealth of flowering trees from areas with subtropical or Mediterranean climates, they can achieve a similar kaleidoscope of color as a canopy for the garden. Instead of planting trees that offer no color except the greens of their leaves, we can select from the material illustrated in the following chapter, the delicate pink of *Albizia julibrissin,* the golden cassias, the blazing red and orange erythrinas, lavender-blue *Jacaranda,* and many others. Trees are the basic, long-lived plants around which we design the rest of the landscape. We must remember that they are not only the providers of shade but can be used to enhance the garden with their flowers and fruits. During their flowering season we are rewarded with a colorful display that adds distinction to the garden and excitement to the landscape.

Flowering trees that delight the tourist in Hawaii or the Caribbean area, such as the golden shower tree, *Cassia fistula,* and royal poinciana, *Delonix regia,* are not adapted to a Mediterranean climate with cool winter rains but may thrive in other subtropical areas, particularly those with heat buildup in summer and summer rains.

This chapter includes some of the more spectacular flowering trees now being successfully cultivated. Some may be best adapted to coastal zones and others to interior valleys. Some will be best in Mediterranean climates, others are more tolerant.

The original sponsors of this book encouraged the use of flowering trees in southern California for many years and the results are now visible in the avenues of *Erythrina caffra* and *Jacaranda,* in commercial sites, and in the Civic Center, Los Angeles, California. Each home gardener can add to the beauty of the landscape by planting one or more flowering trees.

HOW TO GROW FLOWERING TREES AND SHRUBS

Flowering trees and shrubs are no more difficult to grow than non-flowering ones, and their bouquets offer double dividends. Flowering trees, in addition to their jewellike beauty, may make beautiful shade and street trees. Some trees, such as the jacaranda, spread carpets of color on the ground with their fallen petals. The small chore of removing them is a small price for their beauty. Proper placement will eliminate even this problem.

1. Select the best species for the site; never plant a tender plant in cold areas, sun-loving plants in shade, or vice versa.
2. Assure good drainage.
3. Dig a hole considerably larger than the container. For a five-gallon-size plant, mix in one cup of bone meal in the bottom of the hole. Partly fill the hole with topsoil or a mixture of sandy soil and well-rotted compost, never manure.
4. Tamp firmly so ground will not settle after planting.
5. Soak the plant, allow to drain, then remove from can and plant at the same level as the surrounding ground. Fill the balance of the hole with soil mixture, firm down again, and water thoroughly.
6. Keep a basin around the plant at a distance from the trunk. *Never* slope the basin so that the base of the plant is lower than the surrounding ground. To do so invites oak root fungus and other trouble.
7. Deep-water by slow soaking the ground every week or two, depending on weather, size of plant, and soil. Few deeper soakings are much better than frequent shallow waterings. A wilted plant may be revived, but a rotted root is dead!
8. Fertilize only as needed. Many plants from Australia and South Africa resent overfertilizing. Original bone meal usually suffices for first year or longer.
9. Prune for shaping and to remove dead wood. Flowering plants should be pruned after flowering unless fruit is also showy, never just before their blooming season.

ALBIZIA JULIBRISSIN	Leguminosae	10°F (-12°C)
Persian silk tree; mimosa; Constantinople acacia	Summer	Southern Asia

The silk tree, *Albizia julibrissin,* with its soft pink puffs of bloom set off by apple-green fernlike foliage, is a harbinger of summer. The tree grows to 40 feet (12 m) often with multiple trunks that must be pruned if a single trunk is desired. Its somewhat flat-topped shape with horizontal spreading branches provides an inviting canopy of shade in summer and its deciduous nature permits maximum sun during winter. The tropical appearance of this beautiful tree, its graceful form, and the delicate pink glow of the blossoms, make it a worthwhile addition to the landscape as a freestanding specimen or a background accent. Selections are available with deep pink tones to the flowers. In hot, humid areas the tree is subject to mimosa wilt but wilt-resistant types have been selected. Silk tree is widely cultivated, not only in warm temperate but in subtropical areas throughout the world, needing only a sunny location.

Plume albizia, *A. distachya* (*A. lophantha*), with greenish-white spikes is occasionally seen in cultivation. Woman's tongue, *A. lebbeck,* a medium to large-sized tree from North Africa to northern Australia, with cream to greenish-yellow pompons of fragrant flowers, is cultivated in warmer areas. The pods rattle in the wind to give the tree its common name. It is a fast-growing, drought resistant, semideciduous tree used as coffee shade in India.

BAUHINIA BLAKEANA	Leguminosae	25°F (-4°C)
Hong Kong orchid tree	Fall or Winter	Southeast Asia

This tree produces strong color and distinction for autumn and early winter. The rather large, kidney-shaped, gray-green leaves shed partially to display swelling flower buds, produced precociously even on small, young plants. The 6 inch (15 cm) flowers, of butterfly-orchid shape, range in color from cranberry-maroon through rose-purple to orchid-pink, often in the same blossom. In favorable sites the orchid tree may reach 25 feet (7.5 m). It requires good drainage and thrives in full sun or high shifting shade. In some areas it may be subject to chewing larvae. The adaptability of the Hong Kong orchid tree is indicated by its cultivation in such areas as southern Florida, southern California, and Hong Kong.

BAUHINIA VARIEGATA	Leguminosae	25°F (-4°C)
Mountain ebony; purple orchid tree; orchid tree; pink camelsfoot	Spring	India, China

The orchid tree is the most common *Bauhinia* cultivated in the subtropics and tropics. A 20 to 25 foot (6-7.5 m) subject, it is inclined to bushiness or multiple trunks. It also is extremely variable in its blooming period, foliage quality, and leaf-holding characteristics according to soil, exposure, and weather vagaries. It may drop its twin-lobed leaves in midwinter or remain evergreen in moist hot areas. The principal show of orchid-lavender to purple, white, or even pink, broad-petaled flowers is in late winter or early spring, with or without new foliage. Following mild, dry winters, the display is sensational, and blossoms may continue to form intermittently on different sections of the tree, occasionally even into autumn. The plant likes heat but withstands considerable cold. It does best in well-drained sunny locations. Fine specimens occur in many parts of the world—Hong Kong, Costa Rica, Cuba, South Africa, southern California where it occurs from coastal areas inland to Palm Springs, and central and southern Florida.

BRACHYCHITON ACERIFOLIUS	Sterculiaceae	25°F (-4°C)
Australian flame tree; flame bottle tree	Spring-Fall	Australia

This deciduous, fast-growing giant may soar 60 feet (18 m) or more. It is best adapted to warm-summer sections, and blooms more dependably in hotter areas. The shining leaves vary in size and shape, even on one tree, but typically are 5- to 7-lobed and often 10 inches (25 cm) across. Leaves drop before flowers appear, or from unusual cold. The small, almost tubular blossoms are scarlet to orange. They may cover the whole crown or, more often, widely separated parts of it. Frequently, sections of the tree throw copious buds in midwinter or early spring with a hot spell. Just as frequently these buds do not open unless watering is withheld in the spring. The showiest flowering usually occurs in summer, and a mature specimen in a hot location can be an inspiring sight for two to three months. The green, boatlike pods, in heavy clusters, are conspicuous and ornamental. The tree is cultivated in its native Australia as well as in the subtropics throughout the world.

Brachychiton populneus, Kurrajong bottle tree, has small, white, bell-shaped flowers and is a good evergreen tree for low rainfall areas particularly where it has deep soil. A hybrid between Kurrajong and the flame tree has been recommended for Australian gardens.

BRACHYCHITON DISCOLOR	Sterculiaceae	25°F (-4°C)
Pink flame tree; pink sterculia	Summer	Australia

Another fine tree for warm inland areas, the pink flame tree or pink bottle tree is 40 to 90 feet (12-27 m) tall, high-headed, and slightly bottle-trunked. Usually narrowly pyramidal in youth, it becomes more widely spreading in maturity. The 6 inch (15 cm), somewhat maplelike leaves are woolly-white beneath and dark green above. They fall just before the flowers appear. Following sudden cold weather, the entire tree may be bare for a period. Blossoms of rose to pink are backed by short brown wool, which also distinguishes the 6 inch (15 cm) rusty, boat-shaped seedpods. This tree is especially handsome for avenue planting where there is ample accumulated sun heat. Single or grouped specimens may also serve as a focal point or an interesting mass. The tree is best in deep soil in hotter areas.

CALODENDRUM CAPENSE	Rutaceae	25°F (-4°C)
Cape chestnut	Summer through Fall	South and East Africa

The cape chestnut (unrelated to the true chestnut or horse-chestnut) is grown for its profusion of white, rose, and lilac flowers, with brownish purple spots, which are displayed in large clusters from late spring into midsummer. Mature specimens slowly achieve 25 to 40 feet (7.5-12 m) with similar spread. Leaves are medium green and up to 5 inches (12.5 cm) long. The tree is partially evergreen or briefly deciduous, depending on location and season. This tree seldom blooms when young, performing best in rapidly draining deep soil, deeply and infrequently watered. It reaches its climax in areas not far from the tempering ocean. In southern California, Florida, and subtropical Australia, as well as in its native Africa, the tree is known for its spectacular flowering show.

CASTANOSPERMUM AUSTRALE	Leguminosae	25°F (-4°C)
Moreton Bay chestnut; black bean	Summer or Fall	Australia

This shining emerald evergreen is round-headed to almost flat-topped. Seldom more than 25 to 30 feet (7.5-9 m) tall it can attain 60 feet (18 m) under ideal conditions. It is an arresting foliage tree with its large leaves composed of 11 to 15 leathery leaflets up to 6 inches (15 cm) long, remaining attractive throughout the year. Its summer flowers provide an extra dividend. Clear yellow to orange to red, and sometimes two-toned, with long stamens, they occur in 6 inch (15 cm) clusters, even on the trunk and main stems. Cylindrical pods to 9 inches (22.5 cm) add interest in autumn. Slow-growing, suitable for full sun or partial shade, it may grace a lawn or somewhat drier spot. Adaptable to most localities, the tree blooms more abundantly in areas of considerable heat. The Moreton Bay chestnut is cultivated in its native Australia as well as in South Africa, Florida, and California. It has been recommended for gardens in Cuba.

CHIONANTHUS RETUSUS | Oleaceae | -10°F (-23°C)
Chinese fringe tree | Late Spring-Summer | China

Four inch (10 cm) clusters of white fringelike flowers drape branches in late spring. This is one of the most beautiful small deciduous trees growing 20 feet (6 m) high and often as wide. Two to four inch (5-10 cm) ovate leaves unfurl before blossoms appear and in fall the leaves turn vivid yellow. Though possible in all climate zones except desert and oceanside, *C. retusus* flowers best in sandy loam where several frosts occur in winter and summers are hot.

Chionanthus virginicus, white fringe tree or old man's beard, a native of the southeastern United States where it is occasionally seen in cultivation, has been introduced into Australian gardens. It needs moist fertile soils to produce the cluster of white flowers with four narrow, ribbonlike petals followed by blue-black, berrylike fruits in summer.

CHORISIA SPECIOSA
Floss silk tree

Bombacaceae
Summer-Winter

27°F (-3°C)
Brazil

Light orchid pink, purplish-rose, mulberry, and burgundy are color tones that identify the hibiscus-like flowers of the floss silk tree. One color dominates each tree, appearing on the outer ends of the widely separated petals. Usually the petal bases are ivory or white-striped, or spotted brown. These variable 6 inch (15 cm) blossoms burst suddenly from upright, ball-like buds, covering the entire crown, gay and springlike against the dry autumn landscape. Mature specimens provide an unforgettable display of flowers lasting for several weeks.

The tree grows fast the first few years, slowing conveniently before reaching 30 to 60 feet (9-18 m). A lance-straight, grass-green trunk turns gray as the tree matures. Usually it is prominently covered with gray thornlike spines. The light green leaves, with fanlike arrangement of their leaflets, partially or entirely drop from unusual cold, or as bloom appears. The floss silk tree thrives with summer moisture but can be tolerant of adverse conditions when well established.

Chorisia insignis, a smaller-growing tree with a larger bottle-shaped trunk, has white flowers with creamy centers which later turn brown. It is best in drier, well-drained sites.

DOMBEYA CACUMINUM | Byttnerlaceae | 28°F (-2°C)
Early Spring | Malagasy

This superb evergreen tree is rare in gardens and deserves more attention. Trees growing in the Huntington Botanical Gardens in San Marino, California, are 50 feet (15 m) high after 25 years' growth. Plants are narrow in outline, usually with single trunks, and have large maple-like leaves. The hanging clusters of clear rosy pink, 2 inch (5 cm) flowers fall before drying. Flowering begins when plants are mature but if propagated from flowering wood cuttings they will bloom much sooner. The species is being slowly distributed through cuttings and seed from the Huntington plants. This species deserves wider cultivation.

More common and very different is *Dombeya* X *cayeuxii,* pink-ball dombeya or pink snowball, a hybrid often erroneously identified as *D. acutangula* or *D. wallichii.* It is a much-branched shrub or small tree to 30 feet (9 m) with hydrangea-like hanging clusters of smaller pink flowers that appear late in spring and are more frost tender than *D. cacuminum.* After flowering the clusters dry brown and remain hanging on the plant unless removed. It is best to prune this *Dombeya* annually in early summer to prevent top-heavy growth. The United States Department of Agriculture has introduced new selections for Florida gardens.

Other species are occasionally seen in cultivation. *Dombeya spectabilis* from South Africa grows to 30 feet (9 m) and has white blossoms. It withstands temperatures down to 25°F (-4°C). *Dombeya burgessiae* has white fragrant flowers streaked with pink. *Dombeya calantha* is a shrubby species with pink flowers.

GREVILLEA ROBUSTA	Proteaceae	20°F (-7°C)
Silk oak	Summer	Australia

Comblike 6 to 10 inch (15-25 cm) tresses of orange-yellow flowers and coarsely fernlike leaves, deep olive-green above and silver beneath, distinguish the so-called silk oak, which is not an oak and in no way resembles the true oaks. One of the largest exported evergreen trees, 50 to 100 feet (15-30 m), this Australian species was introduced over a century ago into California and is now cultivated around the world as an ornamental in such places as Hong Kong or for coffee shade in East Africa and Costa Rica. Especially appealing as a young nursery plant, it burgeons upward unbelievably fast. Usually pyramidal in youth, its main trunk divides with maturity to develop a curiously free-form, widely variable shape. It is impressively handsome in the background, particularly when flowering full grown. Easily cultivated in any soil and most exposures, dry or wet, its most rewarding display of blossom is usually seen in the warm interior or even in the desert. Its chief faults are very brittle branches which often break in heavy wind, invasive roots, and extensive leaf litter.

HARPULLIA PENDULA	Sapindaceae	27°F (-3°C)
Queensland tulipwood	Fall through Spring	Queensland, Australia

A lavish display of pendulous orange seedpods identifies this comely Australian from fall until summer. Each inflated capsule, an inch and a half (4 cm) across, is two-parted, tangerine or yellow outside, bright red within. Jet black seed peep tentatively from their enclosures to eventually spread the hulls and shine forth. Flowers are inconspicuous. The tree is dark green, round and dense when placed in a sheltered site. Growth is slow and the new tips are tender. Its ultimate height is 45 feet (15 m). Leaves are pinnate usually with two to four pairs of polished leaflets some 6 inches (15 cm) long. In its native habitat tulipwood is valued for its beautifully marked wood used for fine cabinets. Successful culture requires well-drained soil, protection from prevailing wind, and considerable heat in summer and fall.

Closely related and occasional in cultivation are the more tender *H. arborea,* native of India, the Malay Peninsula, and the Philippines, and *H. cupanioides,* native of Indonesia and southeast Asia.

HYMENOSPORUM FLAVUM
Sweetshade; wing-seed tree;
scented yellow-blossom tree

Pittosporaceae
Spring to Summer

25°F (-4°C)
Australia

Honey-scented flowers of soft yellow-orange perfume the environs of the sweetshade from midspring to midsummer. This slender, open evergreen of 20 to 40 feet (6-12 m), with shining light-green 6 inch (15 cm) leaves, which drop with sudden cold or erratic watering, thrives in well-drained soil but sulks in windy places. Its widely spaced limbs and smaller branches may be thickened as well as strengthened by frequent pinching and heading back from extreme youth into maturity. All weak limbs should also be removed. It lends light shade and interesting structure to streets and gardens, flowering for many weeks.

JACARANDA MIMOSIFOLIA | Bignoniaceae | 25°F (-4°C)
Jacaranda | Spring-Summer | Brazil & Argentina

Perhaps the best known and most widely loved of our flowering trees, the jacaranda is cultivated throughout the subtropics and tropics. It is the official city tree of Pretoria, South Africa; lines avenues in Australia, its native Argentina, and southern California; may be seen in Cuba, Hong Kong, Costa Rica, Florida, Hawaii, and many other places. It is distinguished by its great loose clusters of lavender-blue tubular flowers. These smother the entire crown from late spring into late summer, then appear lightly and intermittently, often repeating as late as early winter in warm exposures. Growing rapidly from seed or cuttings, jacaranda reaches 50 feet (15 m) with maturity. The leaf canopy develops when the main show of flowers is gone, to cast welcome shadow in late summer and fall. The fernlike leaves drop in midwinter. This tree is easily adapted to most soils and areas. It becomes dwarfed by prolonged drought, tends to be floppy with constant wetness, and may fail to flower close to the ocean. Flower color and size are variable and selections are available, including a white-flowered cultivar.

KOELREUTERIA ELEGANS | Sapindaceae | 22°F (-5.5°C)
Chinese flame tree, flamegold | Fall | Taiwan, China

This tree is famous for its fall display of salmon seedpods rather than for its earlier brief show of yellow flowers. A moderate grower, 20 to 40 feet (6-12 m), this is an ideal flat-topped patio subject for those who want summer shade and filtered winter sunshine. Shining whorls of golden-green, 1 to 2 foot (0.3-0.6 m), divided leaves, formed of intricate, shallow-toothed, 2 to 4 inch (5-10 cm) leaflets, encircle branch and/twig ends. They turn yellow briefly before dropping, usually in midwinter. Mature trees produce an arresting crown display of papery fruits, each 2 inches (5 cm) long, in flaming salmon masses as if the treetop were wreathed with an odd bougainvillea. Except for the immediate seashore and sites of consistent strong wind, this tree is adaptable to most soils and climates in subtropical areas. It is effective when planted in clumps producing a pleasing accent specimen. Numerous volunteer seedlings may appear under mature trees and are useful for propagation. Chinese flame tree has been in cultivation as *Koelreuteria formosana* and *K. henryi.*

Koelreuteria paniculata, goldenrain tree, varnish tree, China tree, pride-of-India, is more suitable for colder areas. A 30 foot (9 m), roundheaded tree with a blue-green cast to the leaves, it produces bright yellow flowers and yellow to red-brown papery pods. *Koelreuteria bipinnata,* evergreen goldenrain tree or Chinese flame tree, is hardier than *K. elegans* and has somewhat larger similar capsules.

LAGERSTROEMIA INDICA	Lythraceae	10°F (-12°C)
Crape myrtle; pride of India	Summer to early Fall	China

A slow-growing deciduous tree 10 to 25 feet (3-7.5 m) tall, the crape myrtle is best remembered for its late-summer profusion of showy flowers lasting for several weeks in electric rose pink, crimson, burgundy, shell pink, lavender, or white. The petals are crinkled, resembling crepe paper. The smooth slender trunk is especially beautiful, intricately and delicately mottled gray, fawn, and taupe. The winter skeleton is arrestingly strong and clean. Sometimes the leaves turn conspicuously red and gold in late fall. Over 150 cultivars are available in the trade.

Crape myrtle is best for the drier, hotter areas. It mildews near the beach and even inland during some years. The soil should be well drained and deeply watered.

Lagerstroemia speciosa, Queen's crape myrtle or pride-of-India, is a larger tree and more tender. The 3-inch (7.5 cm) flowers are borne in large terminal panicles of pink or mauve and a flowering tree is spectacular. It needs a fertile, acid soil and more moisture than *L. indica.*

MARKHAMIA HILDEBRANDTII | Bignoniaceae | 32°F (0°C)
Markhamia | Summer | Africa

Golden trumpet flowers brighten the dusky crown of this beauty in late summer. The clustered funnelform blossoms, lemon to orange-yellow, sometimes with violet stripes, are slightly suspended in terminal panicles. Bloom may not occur every year. The best show follows a mild, sunny winter. Slow to 30 feet (9 m), the tree is slender with rounded top. Leaves are specially handsome, opposite, pinnately compound to 15 inches (37.5 cm), of burnished olive-green. The new growth is purplish-copper in color. Winter chill further purples the foliage which drops from frost or frigid wind. Best placed in a protected spot with maximum summer heat, this tropical African tree is an interesting addition to the palette of an adventuresome gardener.

MELALEUCA LINARIIFOLIA	Myrtaceae	25°F (-4°C)
Snow-in-summer; flax-leaf paperbark	Summer	Australia

Narrowly upright, open and very rapidly growing during its first several years, the crown of this tree alters with maturity to a roundheaded form of greater density. Thin pale-beige branchlets with clusters of delicately pointed blue-green leaves form an elegant contrast to the honey-brown papery bark peeling on trunk and older limbs.

At all seasons distinguished by its stylized texture and subtle coloring, *M. linariifolia* becomes truly spectacular during early summer in full bloom, giving the effect of new fallen snow. Completely evergreen, this white bottlebrush is happy either in moist lawns or with moderate drought, inland or near the sea. See pages 00, 00, and opposite for other bottlebrush types.

Melaleuca elliptica has deep red blossoms and grows to 12 feet (3.6 m). *M. macronycha* (*M. longicoma*) has orange-red blossoms and grows from 8 to 10 feet (2.4-3 m). *M. radula,* 5 feet (1.5 m), has violet blossoms. *M. steedmanii* has brilliant crimson and gold blossoms and can be used as a ground cover since it tends to remain below 3 feet (1 m). *M. wilsonii* has rosy-purple blossoms and grows to about 5 feet (1.5 m). *M. ericifolia,* a shrub or small tree, tolerates salt spray possibly better than others in this genus.

Melaleuca styphelioides and *M. quinquenervia* (sometimes confused with *M. leucadendron*) are two handsome trees to 40 feet (12 m) with peeling papery tan to whitish bark and creamy white flowers. *M. styphelioides,* prickly paperbark, has pendulous branches and small prickly leaves. *M. quinquenervia,* cajeput tree or paperbark tree, is tolerant of many soils, even taking flooding and salinity. It has larger lanceolate leaves to 3 or 4 inches (7.5-10 cm) long. It has been cultivated widely both in the subtropics and tropics.

METROSIDEROS EXCELSUS	Myrtaceae	25°F (-4°C)
New Zealand Christmas tree	Spring-Summer	New Zealand

The showy, deep red flowers of the New Zealand Christmas tree are dominated by their prominent stiff stamens, numerous and threadlike. These blossoms occur in clusters that submerge the top and sides of this round-headed evergreen from late spring to early summer. The leaves are thick and leathery, shining dark green on young growth, gray and hairy on their undersides with maturity. The tree grows slowly to 30 feet (9 m). By nature it has a multiple trunk, quite bushy in youth and needs occasional thinning to remove dead wood. It is definitely a subject for coastal areas, usually at its best in the salty, equable air of the seashore, where it endures incredible wind and exposure, an ideal focal material for beach gardens. Because of its moderate size and slow growth the New Zealand Christmas tree is useful in small gardens or even in tubs.

Several other species of this genus are occasional in cultivation, differing in flower color and foliage.

See pages 22, 36, 37 for other bottlebrush types.

PAULOWNIA TOMENTOSA	Bignoniaceae	15°F (-9°C)
Royal paulownia; empress tree; princess tree; karri tree	Early Spring	China

Large-leaved, deciduous, and suggestive of catalpa, the empress tree is very old in the history of gardens. It is widely distributed and often naturalized in warmer parts of temperate zones. Forty feet (12 m) or more tall with sometimes greater spread, it grows rapidly and casts deep shade in summer. Surface roots discourage many ground covers. Vigorous limbs branch low from the trunk or trunks and often curve heavily downward. Leaves are broadly ovate to a foot (30 cm) long, roughly heart-shaped or three-lobed, light to vivid green with beige pubescence covering the underside. They scorch in hottest sun. Furry, light brown buds in panicles to one foot (30 cm) high form in the fall, adding interest to the strong-boughed, leafless skeleton. Flowering can be spectacular in early spring unless extremely low temperatures freeze the buds, or near frost causes blossoms to drop before opening. Two inch (5 cm) funnelform flowers, lavender outside, spotted purple and streaked yellow within, crowd the upright panicles. The candelabra-like inflorescence usually precedes new foliage. Persistent 2 inch (5 cm) seed pods, formed the previous season, accompany the flower show. Best placed away from strong wind, paulownia prefers deep sandy loam with good drainage. The tree must be pruned annually to maintain its shape. Paulownia is tolerant of city conditions of smoke and dust and can be used in coastal areas if protected from the wind.

Paulownia fargesii and *P. fortunei* have similar growth, culture, and blooming patterns except flowers tend to be larger and the plants will stand even lower temperatures.

PRUNUS CAMPANULATA	Rosaceae	23°F (-5°C)
Taiwan cherry	Spring	Taiwan and Ryukyu Is.

By far the most successful of the many flowering cherries for subtropical landscapes, this one will blossom reasonably well in sight of the ocean. It apparently has little chilling requirement compared with other flowering cherries and tolerates cool summers. Inland it is spectacular. Just as the flowers of fruiting almond begin to shatter, this cherry bursts into strong color, a penetrating cerise, for several weeks, usually in early spring. The bell-like pendulous flowers are clustered along arching or pendent branches. Leaves are oblong-oval, usually doubly serrate, glabrous above and below. A vigorous tree to some 20 feet (6 m), the Taiwan cherry shows bronze new growth while still in blossom. *Acacia baileyana* is in left background. See page 39.

STENOCARPUS SINUATUS	Proteaceae	27°F (-3°C)
Firewheel or Rotary tree	Fall-Spring	Australia

This remarkable evergreen tree derives its common name from the shape and color of its flower clusters. The 2 to 3 inch (5-7.5 cm) scarlet and yellow flowers are arranged so they resemble a wheel of flame. The Rotary Club of Sydney, Australia, adopted this tree because the wheel-shaped inflorescence resembled the Rotary emblem. Southern California Rotary clubs followed this lead, starting their own plantings of the Rotary tree in public parks. The green buds turn yellow, then open red intermittently throughout the year with usually a peak display in the heat of early fall. The 8 to 12 inch (20-30 cm) rigid leaves are also showy, shiny dark green and leathery, resembling leaves of some evergreen oaks. The tree is rather narrow and often dense, slow growing to 25 feet (7.5 m). It grows best in a well-drained acid soil, and needs deep, slow watering and a somewhat sheltered location with excellent air circulation. Chelated nutrients may be needed to correct occasional leaf yellowing in young plants.

THEVETIA THEVETIOIDES	Apocynaceae	27°F (-3°C)
Giant thevetia	Summer-Fall	Mexico

In cultivation, this plant usually takes the form of a large shrub or shrubby tree, but in the mountains of Mexico between Puebla and Oaxaca it has been seen growing in a dry riverbed as a single-trunk, 15-20 foot (4.5-6 m) tree. It blooms at intervals during the summer and fall with large open trumpetlike flowers in clusters. Its narrow glossy leaves with a pebbly surface persist through the winter.

Thevetia peruviana (*T. neriifolia*), yellow oleander or lucky nut, also grows as a large shrub or small tree with evergreen narrow smooth leaves and smaller yellow or apricot-colored funnel-shaped flowers for about four months during the warm season. The yellow oleander, while not as showy as the giant thevetia, is much more common in cultivation. Native of tropical America it thrives in Hong Kong, Singapore, Cuba, and other subtropical and tropical areas. Both of these species do equally well inland and near the coast. The yellow oleander is well suited to hot desert areas.

TABEBUIA CHRYSOTRICHA | Bignoniaceae | 24°F (-4.4°C)
Golden trumpet tree | Winter or Spring | Brazil

This brown-barked, low-headed and rather open and wide-spreading deciduous tree creates a sensation following a winter cold snap with tight clusters of large golden yellow flowers in dazzling profusion. The mature leaves that follow the blossoms are composed of three to five leaflets of unequal size, dark green and conspicuously veined above, rough-textured pale olive beneath. Twigs and branches also display an interesting detail of matted beige hairs that suggest thin flocking. This elegant tree promises significant ornamentation for moist or slightly dry sites in areas that have the necessary heat buildup. Near the coast cooler temperatures may prevent heavy blooming. Its potential height of 50 feet (15 m) should be considered in planting.

Several other yellow-flowered trumpet trees are in cultivation. One of the more common is the 25 foot (7.5 m) silver trumpet tree or tree-of-gold, *T. caraiba* (*T. argentea*) from Paraguay and Argentina. Showy in flower when leafless, it gives an extra bonus with its silvery leaves. *Tabebuia umbellata,* a 15 foot (4.5 m) tree from Brazil, is adapted to areas with summer rain. *Tabebuia chrysantha, T. guayacan, T. serratifolia, T. rufescens* and *T. spectabilis* are other yellow or golden-flowered trees that may be seen in cultivation.

TABEBUIA IMPETIGINOSA | Bignoniaceae | 24°F (-4.4°C)
(*T. ipe, T. avellanedae*)
Ipe | Spring | Tropical America

This flamboyant trumpet tree has delicate lavender-pink flowers with yellow centers. Blossoms erupt suddenly to cover bare branches in early spring like a host of pastel butterflies. The plant described as *T. avellanedae* var. *paulensis* has smaller flowers of the same color as *T. impetiginosa* but does not grow as tall, and blooms twice a year. Ipe seems best suited to inland areas where there is significant heat buildup.

Other pink or purplish-flowered species in cultivation include *T. rosea* (sometimes erroneously called *T. pentaphylla*), widely grown as amapa rosa, pink tecoma, or roble blanco, a 20 to 60 foot (6-18 m) tree with terminal clusters of 2 to 3 inch (5-7.5 cm) pink, pinkish-purple, or nearly white flowers.

Tabebuia, a genus of some 100 species native to tropical America, is the source of some of the most common showy ornamental trees. They thrive in rich soil and are easy to propagate by cuttings, layering, or seed. They are adapted to full sun but should be protected from strong winds.

Tabebuia chrysotricha

TIPUANA TIPU	Leguminosae	25°F (-4°C)
Tipu tree	Summer	So. Brazil, Argentina

Abundant sprays of apricot to yellow pea-shaped blossoms drip from masses of light green fernlike foliage from early to midsummer. The tipu is hardy and fast-growing, its flat-topped head often spreading wider than its height, which varies from 25 to 30 or more feet (7.5-9 m). Particularly suited to inland summers, this tree furnishes filtered shade for ten months, losing its leaves briefly in midwinter. Its pliable limbs may easily be thinned to produce an extremely open structure, or shortened to achieve greater density and less diameter. It grows easily in most watered soils but is drought tolerant. Tipu makes a noteworthy specimen and it has had wide use as an avenue tree in various parts of the subtropics. It may be seen in many places in southern California as well as along streets in Lima, Peru, and elsewhere.

DELONIX REGIA	Leguminosae	30°F (-1°C)
Flamboyant; flame tree; royal poinciana	Spring-Fall	Malagasy

The flamboyant with its gnarled trunk and branches, umbrella-shaped crown, and fernlike leaves is characteristic of the tropics throughout the world. Native to Malagasy it is one of the several flowering trees that has been called "the most beautiful tree in the world" with its scarlet flowers some 3 inches (7.5 cm) across. The petals are broad at the tip and taper to a claw. The lower petal may be white and yellow with red blotches or two shades of yellow with fewer red blotches. The 1 to 2 foot (30-60 cm) long flat, brown seedpods are conspicuous, particularly when the tree is deciduous in the dry season.

The tree is best in low altitudes near the sea since it is very tender to the slightest frost. Inland in frost-free areas it requires a deep soil and adequate water. It is easily grown from seed. Trees may reach as much as 60 feet (18 m) in height but 20 to 45 feet (6-13.5 m) is more common. The widely spreading crown provides fine shade for avenues.

CHIRANTHODENDRON PENTADACTYLON	Bombacaceae	27°F (-3°C)
Mexican hand plant; monkey-hand tree	Spring-Fall	Mexico and Guatemala

Variously called handflower tree, mano de mico, mano de león, and devil's hand, this tree, one of the most celebrated of Mexican plants, was well known to the ancient Mexicans from a single tree on the slopes of the volcano of Toluca. The several common names describe the flower with five bright red stamens, lined with yellow pollen, tipped with clawlike appendages, and exserted like the fingers of a hand from the large cup-shaped calyx. The brown-woolly, leathery calyx, bright red within, is about 2 inches (5 cm) long. The large lobed leaves of 12 inches (30 cm) long are green above and rusty woolly beneath.

This rapidly growing evergreen tree to 50 feet (15 m) or more is cultivated for its shade and the curious flowers often hidden by the leaves. The tree has been planted widely in central Mexico and coastal southern California. It is best in deep, well-drained soils.

SPATHODEA CAMPANULATA	Bignoniaceae	30°F (-1°C)
African tulip tree;	Intermittent	East Africa
flame of the forest	throughout the year	

The African tulip tree is one of the more striking tropical trees with its brilliant flame-colored flowers. Even the buds are conspicuous—large, pointed, terminal clusters of velvet brown claws that split vertically, two to five at a time, to reveal the wide-belled trumpet flowers to 5 inches (12.5 cm) long, intensely red-orange scalloped and picoteed. Seven to nineteen dark green leaflets 2 to 4 inches (5-10 cm) long and one-and-one-half inches (3.75 cm) wide, conspicuously veined, comprise the pinnately compound leaves to 15 inches (37.5 cm) in length. Rusty-haired young branches and leaf stalks impart a golden brown patina to the foliage canopy.

Native from Guinea to Angola, Sudan and Uganda, African tulip is grown throughout the tropics and subtropics often reaching 50 feet (15 m) or more, evergreen, flowering intermittently throughout the year, producing constant litter. In cooler areas such as Malaga, Funchal, and southern California the tree loses most of its leaves in winter, sulks through the chilly spring, attains glory from late summer into fall. The trees are frost sensitive and should be used only in relatively frost-free areas; however in the tropics they seem to do best in cooler regions. Always advised is deep, well-drained soil away from cold or drying winds.

BOMBAX CEIBA (*Bombax malabaricum; Salmalia malabarica*); Red silk-cotton tree; cotton tree	Bombacaceae	28°F (-2°C)
	Spring & Summer	Tropical Asia

Widely cultivated throughout the world, this large buttressed, spiny-trunked tree may reach 75 feet (22.5 m) in height with a straight central trunk and whorls of horizontally spreading branches. The conspicuous red flowers 2 to 4 inches (5-10 cm) long with numerous stamens appear in the spring before the leaves and are followed by oblong capsules about 6 inches (15 cm) long filled with a soft silky mass of hairs covering small black seeds. The five to seven leaflets are palmately arranged.

Because of its size the tree is not adapted to the home garden but is commonly used in parks and public gardens where its conspicuous shaving-brush flowers attract attention.

Michelia doltsopa

FLOWERING TREES AND SHRUBS

Magnolia

Many outstanding flowering trees belong to the magnolia family. Probably best known are the evergreen and deciduous members of the genus *Magnolia.* Less well known are the evergreen trees in the genera *Michelia* and *Talauma.* Since the members of this family are native to regions of high rainfall, when cultivated in hot inland areas they do best with light afternoon shade with their roots protected by a mulch or ground cover.

Most tolerant of heat and dry conditions as well as salt drift and dune sand is the southern evergreen magnolia, *M. grandiflora,* with large glossy leaves and showy, fragrant flowers; 'Majestic Beauty' is one of its better cultivars. This native of the southeastern United States is widely planted throughout the subtropics. Another evergreen species is *M. delavayi* from China, a rare, handsome-foliaged background tree.

Among the deciduous species that bloom earlier and more spectacularly, the best is *M.* X *soulangeana,* saucer magnolia or Japanese magnolia, with its many cultivars, all large shrubs or small trees covered with white to purplish flowers early in the spring or in late winter. *Magnolia stellata* forms a dense shrub with smaller but profuse white flowers. Especially elegant in form and color of bloom are *M. heptapeta* (*M. denudata*), *M. quinquepeta* (*M. liliflora*) and *M. sargentiana* var. *robusta.*

MICHELIA DOLTSOPA | Magnoliaceae | 25°F (-4°C)
Spring | Asia

Five to seven inch (12.5-17.5 cm) cream-white flowers adorn this 20 foot (6 m) evergreen tree. From fat brown velvet buds, flowers open satinlike and fragrant, suggestive of magnolia. As early as midwinter, they cover the tree, remaining as late as midspring against shining deep green leaves that are thin, leatherlike and 3 to 8 inches (7.5-10 cm) long. Tall in its native Himalayas, it remains somewhat shrubby in the subtropics, upright, and close-limbed. Rich, leafy soil well-aerated, some wind protection and a cool, moist root-run are recommended. Seedlings often appear under the mature trees.

Michelia figo (*M. fuscata*), banana shrub, a native of China, is widely cultivated for its evergreen dark green leaves and small yellow flowers with the odor of banana. In South Africa it is known as port-wine magnolia for its rich aroma. It is best in partial shade in a somewhat acid, fertile, well-drained soil.

THE AUSTRALIAN BOTTLEBRUSHES

Callistemon, Calothamnus, Kunzea, and *Melaleuca* are all known as bottlebrushes. All are members of the Myrtaceae and there are over 100 species in the four genera. Because of their bright color, long blooming period, hardy nature, and the lasting quality of the blossoms in floral arrangements, they are now widely used throughout many regions in the subtropics and tropics. Most are shrub forms though the two illustrated opposite are small trees. Melaleucas are discussed on page 22. *Metrosideros,* the New Zealand Christmas tree, somewhat like the bottlebrushes, is illustrated on page 23.

CALLISTEMON CITRINUS (*C. lanceolatus*)	Myrtaceae	24°F (-4.4°C)
Lemon bottlebrush; red bottlebrush	Winter & other seasons	Australia

Of the many bottlebrushes in cultivation, *C. citrinus* (upper illustration opposite), is most nearly everblooming, and has crimson red flowers. Long treated as a large shrub, this evergreen species is now commonly grown as a small tree. It can be trained into a 20 foot (6 m) tree with a narrow head and a single trunk. The new leaves are pinkish-copper in color becoming vividly green. Usually it blooms conspicuously in midwinter, then intermittently throughout the year. Easily adapted to most adversities of soil, wind, and water, it is also amazingly tolerant of both heat and cold.

Callistemon viminalis, the weeping bottlebrush (opposite lower), is a 20 to 30 foot (6-9 m) evergreen tree with low-sweeping pendent branches softly clothed in narrow, light-green leaves. The flowers hang in dense spikes of pure red, soft and brushlike, encircling the ends of slender branches in brilliant profusion intermittently throughout the year. The tree may break into full display at any season, especially following periods of high temperatures. Afterward it rests for six weeks or so, preparing new growth for another show. The fruits hang on for several seasons, covering the mature branches like carved wooden buttons. The weeping bottlebrush adapts easily to well-drained soils and is the most commonly cultivated throughout the world both in the tropics and subtropics, being recommended for areas as diverse as Florida, southern California, South Africa, Hong Kong, and Cuba. Careful staking and ample water are essential for young plants. Some heading back and severe thinning of surplus branches to prevent top-heavy growth are desirable. Like all *Callistemon* species, iron soil additive may be needed if leaves show yellowing.

Callistemon salignus, also a small tree, has white flowers and its new foliage is coppery-colored. Other desirable bottlebrushes are *Calothamnus quadrifidus* and *C. villosus,* commonly called net bush. These are 6 foot (1.8 m) needle-leafed shrubs with bright red blossoms formed like tufted ends of a bottlebrush. Kunzeas are evergreen shrubs or small trees varying from prostrate species to 20 feet (6 m). Their small brushlike flower clusters are white, yellow, pink, or red.

Acacia baileyana

ACACIA

Leguminosae

The golden glory of acacia trees blooming from midwinter through spring and then sporadically throughout the year invariably attracts admiration because of their airy beauty.

There are approximately 800 species of *Acacia* described from the warmer parts of the world, about half of them from Australia. Of these, probably fewer than 100 have been used in gardens and these largely in areas with a Mediterranean climate of warm, dry summers and cool, wet winters. These species from dry areas are well adapted to poor soil, little water, and less care. They languish in lawns or watered gardens. There are species, however, that will take garden conditions, some are best in partial shade and some will take heavy clay. Most of the acacias that are available are frost tolerant but the majestic *A. terminalis* (*A. elata*) is tender when young.

Acacias are said to have two serious faults. They are relatively short-lived and their branches are brittle and easily broken by the wind, both faults that can be mitigated by proper culture. An acacia planted in the location to which it is best adapted, watered to encourage deep rooting, and pruned and properly staked to strengthen its trunk will often outlast its normal life expectancy.

The low shrubby acacias that are most promising for the average home garden have hardly been tried. Their small size, compact shape, and long blooming periods make them desirable and many of them adapt well to garden conditions. A few species are available at nurseries. The keen gardener who wants to try some of the others can easily grow them from seed, using the boiling water method to break down the hard seed coat. Many of them will grow rapidly and bloom in 18 months.

Acacia glandulicarpa

ACACIA BAILEYANA	Leguminosae	20°F (-7°C)
Silver wattle; fern-leaf acacia;		
Cootamundra wattle;		
Bailey's acacia	Winter	Australia

This is a fast-growing 20 to 30 foot (6-9 m) tree for rather dry slopes and well-drained soil. With the New Year *Acacia baileyana* bursts into a fluffy cloud of clear yellow fragrant flowers displayed against the blue-gray head of fernlike evergreen foliage. A distinguished, much rarer variation with purple-tinged foliage is *A. baileyana* 'Purpurea.'

Acacia glandulicarpa, one of the best shrub acacias, is a compact mass of green most of the year. In spring it turns into a golden ball seven to eight feet (2.1-2.4 m) tall and broad with tiny puff-ball, delightfully fragrant flower clusters. This species as well as *A. lineata, A. gladiiformis, A. cardiophylla, A. cultriformis* and others are useful shrubs for slopes or freeway embankments.

ACACIAS CAN PROVIDE COLOR ALL YEAR

With judicious selection of species, one may have acacia bloom every month of the year in warmer areas. It should be kept in mind, however, that many acacias bloom erratically or intermittently throughout the year, such as *Acacia retinodes,* sometimes sold as *A. floribunda,* a 20 foot (6 m) tree that is almost everblooming in coastal areas.

For winter flowering the following are recommended: *Acacia alata,* a 5 foot (1.5 m) leafless shrub with pale yellow flowers; *A. calamifolia,* an 8-12 foot (2.4-3.6 m) shrub with needlelike leaves and bright yellow flowers; *A. cardiophylla,* a 10-16 foot (3-4.5 m) shrub with arching branches and bright yellow flowers; *A. cultriformis,* a shrub to 15 feet (4.5 m) with silvery gray leaves and yellow flowers; *A. dealbata,* a 50 foot (15 m) tree with yellow-gold fragrant flowers; *A. decurrens,* a 40 foot (12 m) tree with fragrant bright yellow flowers; *A. podalyriifolia,* a 20 foot (6 m) shrub or small tree with silvery gray leaves and light yellow fragrant flowers; and *A. rupicola,* a five foot (1.5 m) shrub with fragrant showy flowers throughout the winter.

Peak of acacia flowering is in the spring with *Acacia acinacea,* a 3 to 6 foot shrub with pale yellow showy flowers; *A. cyanophylla,* a 20 foot (6 m) tree with bluish leaves and showy orange flowers.; *A. mearnsii,* a 40 foot (12 m) tree with creamy fragrant flowers that may continue intermittently throughout the summer; *A. pendula,* a 15 foot (4.5 m) tree with feathery leaves, graceful, drooping branches and fragrant flowers; *A. pycnantha,* a 25 foot (7.5 m) tree with light green leaves and golden yellow flowers; *A. riceana,* a 20 foot (6 m) tree with pendulous branches and pale yellow flowers; *A. saligna,* a showy 20 foot (6 m) tree with pendulous branches, vivid yellow flowers; *A. subporosa,* a 40 foot (12 m) tree with pendent chains of fragrant whitish flowers; *A. verticillata,* a 20 foot (6 m) conifer-like shrub or tree, with pale yellow flower spikes; *A. vestita,* a 10 foot (3 m) shrub or tree with pendulous branches, ashen gray leaves.

Summer bloomers include *A. karroo,* a South African 20 foot (6 m) tree with rich green leaves and yellow flowers; *A. obliqua,* a 4 foot (1.2 m) shrub with arching branches and yellow flowers that may begin flowering in mid-spring and continue to midsummer; *A. pruinosa,* a 60 foot (18 m) tree with coppery new growth and pale yellow flowers; *A. woodii,* a South African 30 foot (9 m) flat-topped tree with pale cream flowers; and *A. terminalis* (*A. elata*), a 60 foot (18 m) tree with dark-green leaves and creamy flowers.

Autumn flowerers include *A. alata, A. mearnsii, A. retinodes,* and *A. rupicola.*

CASSIA SPECTABILIS	Leguminosae	26°F (-3°C)
Crown of gold tree	Fall	Argentina

Pictured above is *Cassia spectabilis,* a tree that has been erroneously called *C. carnaval,* closely related to *C. excelsa.* This partially evergreen tree displays abundant 12 to 16 inch (30-40 cm) upthrust spikes of bright yellow flowers at the ends of the light green branch tips, usually in very late summer or early fall. It grows rapidly to 40 feet (12 m) or more and thrives in full sun with fairly fast-draining, warm soil, in somewhat wind-protected locations. It is tender when young, but has tolerated 26°F (-3°C) when mature. It needs moisture during the growing season and should be pruned hard after flowering.

Cassia multijuga from Brazil is an upright tree with smaller leaves and very long panicles of clear yellow flowers that appear in the early fall. *Cassia liebmanii* from Mexico is a pagoda-shaped tree to 20 feet (6 m), blooming twice a year with flowers somewhat smaller than *C. spectabilis.*

Cassia leptophylla (above), the gold medallion tree from Brazil, blooms in the summer, and has exceptional beauty. It is open-headed, low-spreading and near evergreen with fernlike foliage. Branch ends produce a spectacle of 6 to 8 inch (15-20 cm) golden yellow clusters, brilliant for several weeks. This species is one of the most shapely and graceful of the cassias in cultivation. Like all other cassias, it grows fast to its mature height of 30 feet (9 m).

Many cassias are attractive shrubs. Among the most popular is *Cassia artemisioides* (below) a silvery, almost needle-leafed shrub from Australia with bright buttercup yellow flowers. It attains a height of 4 to 6 feet (1.2-1.8 m), needs good drainage, and blooms several times a year. It is useful for the small garden and grows well in desert areas. It must not be overwatered. *Cassia surattensis* var. *suffruticosa,* native to tropical Asia and Polynesia, is one of the most prolific blooming shrubs, providing a show of color from early summer into late fall and forming a dense mass to 8 feet (2.4 m) high and about as wide. When grown in rich sandy loam in full sun or very light shade and given ample moisture, it offers a very lush green tropical appearance. In cooler areas it may become partially deciduous and will suffer frost damage below 23°F (-5°C). Plants will flower the second or third year from seed. *Cassia coquimbensis* is a 12 foot (3.6 m) shrub from Chile which blooms in spring and again in fall, with dense clusters of one and a half inch (3.75 cm) flowers. Other desirable shrub cassias are *C. alata* (tender), *C. bicapsularis* (tall and winter blooming), *C. eremophila* (a dry grower resembling *C. artemisioides*), *C. corymbosa, C. helmsii, C. biflora, C. chatelainiana* and *C. didymobotrya.*

THE REMARKABLE EUCALYPTUS

Myrtaceae

Landscapes in many parts of the world would look strange indeed without the familiar eucalyptus, introduced in large numbers from Australia in the nineteenth century.

The genus, a member of the myrtle family, comprises about 500 species mostly native to Australia. All are evergreen and characterized by a bud-cap or operculum that pops off to reveal the beauty of a flower made up mostly of showy stamens. Species vary from 5 foot (1.5 m) shrubs to forest giants of 300 feet (90 m) or more occurring in such varied climatic conditions as wet forests, winter snows in high mountains, deserts, and Mediterranean-type climates.

When most people hear the word *eucalyptus* they are apt to think of the most common species introduced throughout the world, the Tasmanian blue gum, *Eucalyptus globulus,* with its giant 180 to 200 foot (30-60 m) stature, ever-shedding bark, and greedy roots. This one is useful as a windbreak but its white flowers are not particularly showy and it is not suitable for urban landscaping. It has been planted in many areas for firewood as well as windbreaks. If a tall eucalyptus is desired, *E. citriodora,* the lemon gum, with smooth, white trunks and lemon-scented leaves is recommended. Its small, white flowers are inconspicuous but the tree is a valuable addition to the landscape with its striking bark color and airy crown form.

This book, however, is concerned primarily with those species of eucalyptus having showy flowers, many of which are native to the semiarid regions of southwestern Australia. Most are small in stature, bloom over periods of two to three months or more and sometimes several times a year. Flowers come in a wide range of colors including red, pink, yellow, orange, green, cream, and mauve. The blossoms make showy floral arrangements, keeping for a week or more in water when the cut ends of the stems are split. Many of these colorful species are recent introductions and have not yet received the attention they deserve from the nursery or floral industries, landscape architects, and home owners. These plants are not "messy" and make ideal small trees for the home garden. Their cultivation is simple and most of them will tolerate average garden conditions provided they have good drainage and are not overwatered. They thrive in both coastal and inland areas and will tolerate temperatures down to the midtwenties, except for *E. ficifolia* which may be injured by temperatures below 27°F (-3°C).

It is difficult to place some *Eucalyptus* species into tree or shrub categories. Many are intermediate. In addition to the four species described and illustrated on pages 48, 49, 50, and 51, the following are worthy of attention.

Species to be planted for winter flowering are *E. preissiana,* a 10 to 15 foot (3-4.5 m) shrub or small tree with large yellow flowers; *E.* X 'Torwood,' a natural hybrid of *E. torquata* and *E. woodwardii,* originating in Western Australia, a 10 to 15 foot (3-4.5 m) tree with dark yellow or orange-colored flowers; *E. tetraptera,* a shrubby tree with flowers having a large four-sided red flower base and red stamens; and *E. pyriformis,* a shrubby tree with large flowers in colors varying from red or pink to cream.

Eucalyptus rhodantha

Eucalyptus X orpetii

Some species bloom not only primarily in winter but at intervals throughout the year. Among these are *E. forrestiana* and *E. stoatei,* both small trees with flowers having a red base and yellow stamens, resembling small Chinese lanterns; *E. kruseana* reaches 20-30 feet (6-9 m) and has pale yellow, almost white flowers and small round silvery leaves; and *E. orbifolia,* a 15 to 20 foot (4.5-6 m) shrubby tree with round silvery leaves and clusters of creamy-yellow flowers.

Winter and spring flowering eucalyptus include *E. caesia,* a 15 to 20 foot (4.5-6 m) tree with pink flowers; *E. macrocarpa,* a 4 to 6 foot (1.2-1.8 m) shrub with large silvery leaves and large 3 inch (7.5 cm) red flowers; a plant listed in the California trade as *E.* X. *orpetii,* presumably a natural hybrid between *E. caesia* and *E. macrocarpa,* which shows considerable variation in the seedlings as to flower size and color as well as leaf shape and color, with all the variants attractive shrubs; and *E. megacornuta,* a 15 to 25 foot (4.5-7.5 m) tree with a single trunk and clusters of large green flowers.

A summer flowering species, *E. calophylla rosea,* resembling *E. ficifolia* (page 50) in size, shape, and flower color, is attractive also for its bronze-pink new foliage. *Eucalyptus macrandra,* a 15 to 20 foot (4.5-6 m) tree with single or multiple trunks, bears clusters of small yellow flowers from summer into fall. If the cut flowers are dried before the stamens fall, they make attractive dry arrangements. *Eucalyptus woodwardii,* 10 to 15 feet (3-4.5 m) tall and somewhat rangy with silvery leaves and bright yellow flowers, blooms primarily in fall and winter.

Some species of *Eucalyptus* flower intermittently and for prolonged periods throughout the year, and flowers may be found in almost any month. Among these are *E. rhodantha* (page 45), a 4 to 8 foot (1.2-2.4 m) shrub with large round leaves and red flowers; *E. torquata,* a 10 to 15 foot (3-4.5 m) tree with showy clusters of small coral-red flowers; and *E.* X 'Helen Ayres,' developed in California by Helen Ayres, a hybrid between *E. rhodantha* and *E. woodwardii.* One specimen, after seven years, attained a height of about 15 feet (4.5 m). It has short-stemmed oval leaves and large red flowers. Possibly other seedlings of this cross will show interesting variations in flower color and leaf structure.

EUCALYPTUS SIDEROXYLON | Myrtaceae | 20°F (-7°C)
Red ironbark | Winter or Spring | Australia

There are two forms of this eucalyptus. One is quite pendulous, often weeping to the ground at maturity; the other is more upright, its foliage slightly greener. Both of these evergreens can attain 40 feet (12 m) and have persistent red-brown bark and gray-green leaves that partially color mahogany in cold weather. The tough light branches drape themselves with skeins of fragile rose-pink to red blossoms that appear intermittently from autumn to spring. Both grow fast in a wide range of soils and microclimates from beach to desert. They are resistant to most wind with little pruning, and are seldom subject to disease or insects.

EUCALYPTUS ERYTHROCORYS Red cap gum	Myrtaceae Fall and at intervals	25°F (-4°C) Western Australia

The 3 inch (7.5 cm) chartreuse to golden-yellow flowers of this eucalyptus are borne in heavy clusters and emerge from buds covered with a bright scarlet cap. Its polished bright green leaves are long, rather thick, and variably shaped. The seed capsules are heavy and attractive. Unlike many species of eucalyptus, *E. erythrocorys* tolerates summer watering where there is reasonable drainage.

This eucalyptus may be grown with either single or multiple trunks and can attain 30 feet (9 m). It is clean, hardy, and fast-growing, blooming a month or more at a time at least three times a year. The cut flowers and buds are effective in arrangements and last well.

EUCALYPTUS FICIFOLIA Flame eucalyptus	Myrtaceae Summer or any season	28°F (-2°C) Western Australia

This eucalyptus varies in blossom color from pure crimson to scarlet, orange, salmon, pink, white, and yellow-green. Selected color forms are available. The flowers average 2 inches (5 cm) across and occur in large clusters like tight bouquets, rather evenly spaced over the rounded crown of leathery dark green leaves. The bark is dark, furrowed, and persistent. The tree grows slowly to 35 feet (10.5 m) with similar spread. The cut flowers are effective indoors, as are the large gray-brown woody seedpods. At its best within a few miles of the ocean, *E. ficifolia* dislikes wet, slowly draining soil. It is a favorite subject for lining avenues.

EUCALYPTUS LEUCOXYLON MACROCARPA 'ROSEA'	Myrtaceae	22°F (-5.5°C)
Dwarf white ironbark	Summer and Fall	Australia

This brilliant-flowered variety to 25 feet (7.5 m) is smaller than the long-cultivated species known as white ironbark. An attractive white trunk and long pointed leaves add to its charm. Free and variable in form, it retains the subtle coloring of its better known relative, but is more widely adapted to urban planting and far more showy in flower. The tree may bloom when only three years old, the abundantly massed flowers a clear, deep red. Easily at home in most soils, it usually blooms well either inland or near the coast.

Erythrina coralloides

ERYTHRINA

The Coral Trees and Shrubs

Leguminosae

Of all the flowering trees, the coral trees, or erythrinas, are among the most spectacular. These trees have been ornamentals in Mexico since Aztec days and have been adopted as the official flowering tree of the city of Los Angeles, California, reflecting its colorful Mexican heritage.

Coral trees are members of the pea family (Leguminosae) and have pea-like blossoms in various modified forms. The hundred or more erythrinas are native of relatively cool tropical and subtropical areas in all continents. Some coral trees are best for mild coastal areas, while others are hardy in inland areas. Some are shrubby, others are large shade trees of 40 feet (12 m) or more. Some are suitable as street trees, others for parkway planting or as

Erythrina X *bidwillii*

landscape specimens. Some are evergreen, others are deciduous with an interesting structure of bare branches.

The genus name *Erythrina* is from the Greek for red, but flowers vary from pink through crimson, scarlet, and burnt orange. The blooming period lasts four to six weeks or longer, and different species bloom at different seasons.

Perhaps the most colorful and one of the hardier coral trees is *Erythrina coralloides* (naked coral or cone flame tree), native of Mexico (illustrated on page 52). The blossoms are clustered on the ends of branches like crimson candles. There is also a pink form. For about six weeks in the spring the tree is without leaves. It seldom exceeds 20 feet (6 m) and the branches bend down, requiring judicious pruning.

The shrubby corals include *Erythrina* X *bidwillii,* a hybrid of *E. crista-galli* and *E. herbacea* (Florida coral bean). It is an exceedingly handsome plant attaining a height of 8 feet (2.4 m), rarely to 20 feet (6 m) or more, is useful for the small garden, withstands severe frost, and can bloom from early spring to late fall. The spikes of red flowers should be cut back after blooming.

Erythrina humeana, Natal coral (picture on page 54), a native of South Africa, produces red-orange spikes of penetrating brilliance terminating the branches while this coral is still a 3 foot (0.9 m) sprout. The flowers are primarily produced in the autumn but in a hot location specimens have been known to bloom almost constantly and copiously from early summer to midwinter. Natal coral may reach 30 feet (9 m) in height and it is suggested that multiple trunks trained to staggered heights can keep many flowers in easy view. This is one of the deciduous erythrinas and the branches are bare for two or more of the winter months, but it possesses interesting structure. It always requires well-drained soil and in cooler areas should be placed in protected positions. In Africa the shrub form is sometimes designated *E. humeana* var. *raja.*

Erythrina humeana

Clusters of brilliant red sickle-shaped flowers in the spring and moderate frost tolerance make *Erythrina falcata* (illustrated page 55), a large evergreen tree native of Peru, a valuable addition to the landscape. Although it may not bloom for 15 years or more when grown from seed, it blooms much sooner when grown from cuttings of wood from mature trees. The tree has an upright habit of growth and may attain a height of at least 50 feet (15 m), making it a striking subject for specimen plantings. A pink-flowered variant is occasionally found.

Erythrina lysistemon (kaffirboom) illustrated on page 56, is another valuable introduction from South Africa. An almost evergreen tree with rounded crown, it may grow to 40 feet (12 m). Slender red blossoms fold back against the stem in late winter and early spring while the tree is essentially leafless. Kaffirboom is frost-tender and best suited to coastal areas.

Erythrina caffra (coast kaffirboom), a native of South Africa, can reach 80 feet (24 m) and provide a spectacular display in parkway plantings. This large, spreading tree with usually flattened crown is not suited to small gardens. The trees are deciduous for a short time before flowering and the bare branch structure is attractive. Clusters of large, wide-open, scarlet flowers appear in late winter and early spring. The tree is resistant to oak root fungus (*Armillaria mellea*).

Erythrina americana (colorín), native of Mexico, long cultivated in Europe, Hawaii, and the West Indies, is a small upright tree to about 25 feet (7.5 m) and hardy to at least 25°F (-4°C). In Mexico City it is planted as a street tree, and is common on the campus of the University of Mexico. The tree is usually deciduous in winter, although in protected areas near the ocean it may be evergreen. The red flowers appear in the spring, and are sold in Latin American markets for salad.

Erythrina falcata

Another hardy, deciduous coral tree is *Erythrina crista-galli* (cockscomb tree or cockspur coral tree) from Brazil. It may bloom in spring, fall, and irregularly during summer, with showy erect, pink to red flowers. The flowering branches die back after blooming. They may be pruned to the stubby trunk of the 15 foot (4.5 m) tree.

Erythrina acanthocarpa (tambookie thorn), native of South Africa, is a spiny bush about 4 feet (1.2 m) high and well adapted to the small garden. This deciduous species tolerates considerable cold. Its scarlet flowers with greenish-yellow tips are breathtaking for three or four weeks in spring.

Erythrina tahitensis (*E. sandwicensis*), a 25 to 30 foot (7.5-9 m) tree from the South Pacific, is somewhat frost-tender. Showy masses of red to salmon-colored, sometimes white, flowers appear in spring. This tree is probably limited to the immediate warm coastal areas.

Erythrina latissima from South Africa, a small tree with very corky bark, is occasional in cultivation but not as showy in flower as some of the other species.

Erythrina vespertilio, batswing coral tree, native of Australia, is another ornamental useful only for warm coastal climates. This tree can attain 25 feet (7.5 m) and its scarlet flowers and curious heart-shaped leaflets are attractive.

Erythrina lysistemon

Coral trees are easy to grow, rooting readily from cuttings, and require little attention. Most important is selection of the species hardy in your area. Most coral trees are rapid growers and space must be allowed for the often spreading, rounded crowns of the tree species. They should not be overwatered and, as with most other plants, good drainage is important. The trees flower best if deliberately dried off for a short period before formation of flower buds. Any pruning necessary to shape the crown should be done after blooming to avoid cutting away the next year's flower buds. Exceptions to the little pruning rule are *Erythrina crista-galli,* which dies back to its main branches in the fall, with blossoms forming on new wood; and *E.* X *bidwillii,* which forms new flowering branches each spring.

Many other species as well as presumed hybrids of coral trees are in cultivation around the world and the genus deserves wider planting and testing for adaptability in all subtropical and tropical areas.

II
FLOWERING SHRUBS

In subtropical and tropical areas it is possible to have shrubs in bloom throughout the year. Many gardens have room for only a few trees but most can accommodate a variety of shrubs to provide continuous displays of color. It may take several years for a tree to bloom but most shrubs will produce flowers while still young. For this reason and because they are so significant in landscape use they have been subjects of hybridization and selection throughout the ages. Camellias have been grown for their beauty for over 900 years; the Persians had gardens of roses; the ancient Greeks used roses in their wreaths; and in the Roman Republic roses, not medals, were awarded to military heros. It is little wonder that we now grow hundreds of cultivars of these longtime favorites.

The choice of shrubs is even greater than for trees. The following pages include the true shrubs, woody plants that produce shoots or stems from the base and do not have a single trunk. Many shrub species of *Acacia, Eucalyptus, Magnolia, Cassia,* and *Erythrina* are discussed in the preceding chapter. It is not easy to tell a tree from a shrub since some shrubs, with pruning, can make satisfactory small trees and several small trees planted together can produce a shrub effect. Many nonshrubby plants such as strelitzia, poinsettia, aloe, and many succulents can be used in the same way as shrubs. All these are included in other chapters in this book. Shrubs offer an additional bonus since many of them make ideal cut materials for indoor decorations; others can be grown satisfactorily in containers for the patio.

The final selections for this chapter, while they represent only a fraction of the desirable shrubs suitable for subtropical gardens, are most significant for the wealth of color they can add to the landscape. They have been chosen for their ability to grow easily under average garden conditions in subtropical climates.

Many additional shrubs are described in the chapter on California native plants and others are listed in Appendix 2. It is impossible in the space available to include all the cultivars and species that may be in the trade. The gardener will find other shrubs in the nurseries and to obtain the flower color or form desired it is recommended that the selections be made when the plants are in bloom.

Rhododendron species and hybrids

RHODODENDRON | Ericaceae | Tender to hardy
Rhododendron; azalea | Spring | Asia, North America

Though shrubs commonly called rhododendrons and azaleas both belong to the genus *Rhododendron,* most people refer to the large 8 to 10 foot (2.4-3 m) evergreen shrubs with 5 to 6 inch (12.5-15 cm) deep green leaves when they use the term rhododendron, and to the 2 to 4 foot (0.6-1.2 m) evergreen or deciduous forms with smaller leaves and flowers when they use the term azalea. The genus *Rhododendron* contains more than 500 species and innumerable cultivars varying from tiny alpine tufts native to the windswept barrens of Tibet to huge rain forest trees 80 feet (24 m) tall with shining 32 inch (80 cm) deep green leaves. Native over a wide area they thus show a wide variation as to temperature tolerance.

A few cultivars of the large-leaved group, such as 'Pink Pearl,' are adapted to the subtropics. The popularity of the smaller leaved azaleas, covered as they often are with a complete mantle of blossoms (see page 59), has increased rapidly. Since they do well in either sun or shade the southern indicas or sun azaleas are used most frequently. Popular cultivars are: 'Pride of Dorking' (cerise red); 'Formosa' (lavender purple); 'George Lindley Taber' (orchid pink); 'Southern Charm' (pink); and 'Glory of Sunninghill' (orange). The fragrant white cultivar, 'Alaska,' is almost everblooming and needs partial shade. The kurumes do equally well but their flowers are smaller. Less well known are the satsuki azaleas, distinctive in their blooming period from late spring to early summer, very large and variably colored flowers, and relatively low, compact growth. Some of these are 'Gumpo' (white); 'Pink Gumpo' (rose-pink); 'Gunbi' (large single light pink); 'Geisha' (single rose-red); 'Kagetsu Muji' (large, single white); 'Shinnyo-No-Tsuki' (magenta with white centers); and 'Eikan' (white, variegated pink).

Recently species of the section *Vireya,* popularly called Malesian rhododendrons, have been introduced. These show great promise for areas with warm summers and frost-free winters, for they may be damaged below 28°F (-2°C) but withstand summer heat if lightly shaded. They are unusual in flower shape, often startling in their colors, and may bloom several times yearly. Some successfully grown outdoors are: *R. laetum* and *R. aurigeranum,* with clear yellow flowers; *R. zoelleri,* a striking orange-yellow; and *R. jasminiflorum,* with tubular white flowers reminiscent of *Daphne.* All but the last are from New Guinea. Two excellent hybrids of this group are *R. phaeopeplum* X *R. lochae,* with fragrant flowers of an unusual pink, and *R. wrightianum* X *R. lochae,* with compact growth and small, pendent, ruby-red tubular flowers.

For all types of *Rhododendron,* planting soil must be loose, high in humus, and slightly acid. Good drainage is a must. Some growers recommend planting in raised beds of nursery soil mix to assure good drainage, and the use of an insulating mulch. With alkaline soils and water, iron deficiency chlorosis can be a problem unless iron chelates are used. Occasional application of soil sulfur will help maintain the proper soil acidity.

BRUNFELSIA PAUCIFLORA 'FLORIBUNDA'	Solanaceae	27°F (-3°C)
Yesterday, today, and tomorrow	Spring-Summer	Brazil

The common name is suggested by the fact that the 2 inch (5 cm) flowers open rich violet, fade to a blue-lavender, and finally become white, so the three shades are always seen on this shrub at the same time. *Brunfelsia* grows to 6 feet (1.8 m) with nearly evergreen glossy leaves. It is somewhat tender but grows equally well in coastal or inland areas in many parts of the world from Florida and California to Singapore and Durban. If in a very hot, dry location it appreciates part shade; it responds to rich, loose compost, liberal feeding during the growing season, and an adequate amount of water.

Brunfelsia pauciflora 'Macrantha' is slightly more spreading in form and has larger leaves and flowers to 4 inches (10 cm) across. *Brunfelsia pauciflora* 'Eximia' has smaller leaves and blooms profusely over a shorter period in spring. *Brunfelsia americana,* with creamy, fragrant flowers to 4 inches (10 cm) across, can be grown in the warmest gardens.

CALLIANDRA HAEMATOCEPHALA	Leguminosae	29°F (-1.6°C)
(*C. inequilatera*)		
Pink powder-puff bush [*left*]	Winter	Bolivia

Somewhat tender, evergreen, probably at its best in the milder coastal areas, this aristocratic shrub also performs well in some inland areas if given protection against a south- or west-facing wall. It bears large rose-pink, powder-puff-like flower balls with showy colored stamens from fall to early spring. It can be grown as a rounded or free-form shrub, 6 to 8 feet (1.8-2.4 m) tall, allowed to tumble over a wall, or espaliered against a wall, for which it is admirably suited. There is also a beautiful white variety. A rainbow calliandra is available with the new growth mottled with gold.

In the American tropics where the genus is native many species have been brought into local cultivation but are not generally available in the commercial trade.

CALLIANDRA TWEEDII	Leguminosae	25°F (-4°C)
Trinidad flame bush [*right*]	Spring and Fall	Brazil

Trinidad flame bush is somewhat hardier than *C. haematocephala.* Its small, green, fernlike leaves tinged red when young, make a pleasing contrast to the brilliant clusters of crimson stamens. It seldom grows higher than 8 feet (2.4 m) with an equal spread, and makes a colorful accent in the garden either in coastal areas or inland. It is fairly drought resistant but grows well under ordinary garden conditions provided it has very good drainage.

Calliandra surinamensis is frequent in cultivation in warmer areas. Its widely arching branches are covered with small flower clusters of white stamens, tipped pink.

CARISSA MACROCARPA (*C. grandiflora*) Natal plum — Apocynaceae — 26°F (-3°C) — Spring through Fall — South Africa

This handsome evergreen shrub is noteworthy both for its 1 inch (2.5 cm) white flowers with orange blossomlike fragrance and its reddish small plumlike edible fruits, often appearing at the same time. The fruit can be eaten raw or cooked, the flavor resembling that of cranberries, or used to make an excellent jelly. All parts of the plant, including the fruit, contain a milky juice. The shrub grows to 18 feet (5.4 m) and does equally well in coastal and inland areas. It is tolerant of sand and salt. It is much branched and bears spines, making it a good hedge plant. Many cultivars are available, some dwarf, compact plants with larger flowers, others lacking spines.

CAMELLIA SASANQUA 'YULETIDE' | Theaceae | 15°F (-9°C)
Sasanqua camellia | Fall-Winter | China, Japan

Tidy appearance, profusion of colorful flowers during cool months, and relative cold-hardiness explain the popularity of camellias in the subtropics. They are relatives of the tea plant, and legend has it that they came to England when shrewd Chinese merchants, to retain their monopoly on the tea trade, substituted camellias for tea plants when an English horticulturist tried to import live tea plants from China.

Camellia sasanqua is one of the three principal species in cultivation, the others being *C. japonica* and *C. reticulata.* It is a graceful, small to medium-sized shrub with glossy evergreen leaves and 3 inch (7.5 cm) flowers in various shades of red, pink, and white. Unlike most species, *C. sasanqua* does well even in almost full sun.

Camellia japonica is the most commonly available species. *Camellia reticulata* has the largest and showiest flowers though foliage is more sparse and the leggy stems call for lower shrubs such as azaleas or other camellias to be grown in front of them. Hybrids of *C. reticulata* with *C. japonica* combine the fine garden qualities of both species and doubtless many cultivars of this type will continue to be introduced. Because of the many cultivars available, varying in flower form and color, selections should be made from flowering plants.

Intriguing *C.* X *williamsii* hybrids have come rapidly to prominence because of their adaptability to both sun and cold. These are crosses of *C. saluenensis* and *C. japonica.* One of the first of these, and still among the best, is *C.* X *williamsii* 'J. C. Williams.'

The foliage of camellias varies considerably but often is so handsome in floral arrangements that it is harvested for florist use. Pruning and disbudding will increase the quality of flowers, and sometimes the plants are treated with gibberellin to further increase flower size. A few have a delicate spicy fragrance and breeders are working to increase this characteristic. Camellias appreciate good drainage, slightly acid soil, should not be allowed to dry out, and are intolerant of soil or mulch accumulating about the base of the trunk. Occasional application of soil sulfur will be helpful to maintain soil acidity. Also acid residue fertilizers should be preferred.

CHAMELAUCIUM UNCINATUM	Myrtaceae	27°F (-3°C)
Geraldton waxflower	Winter-Spring	Western Australia

Useful both in the garden and for cut flowers, this 6 foot (1.8 m) shrub with fine foliage and delicate sprays of flowers in varying shades of pink is one of the finest contributions of Australia to the landscape and cut sprays are used by commercial florists as fillers. The plant blooms for at least three months and is benefited by heavy pruning after flowering. It should be cut back approximately one-third of its height each year for maximum bloom. It needs good drainage, does not tolerate manure, and is equally happy near the coast or inland.

CISTUS 'DORIS HIBBERSON'	Cistaceae	20°F (-7°C)
Rockrose	Spring	Mediterranean

The discovery that these colorful, tough, hardy, evergreen, low-growing shrubs from the Mediterranean area are also fire-resistant has focused attention upon them in southern California during recent years. The Los Angeles State and County Arboretum, after conducting numerous experiments, has concluded that these plants may char when exposed to intense heat but do not burst into flame and has suggested their suitability for planting along firebreaks and in foothill areas. This and other *Cistus* species are quite drought tolerant, and mature plants should be watered with care to avoid root-rot.

The plant illustrated has 3 inch (7.5 cm) flowers. Other species have flowers that are white, lilac, or purple. *Cistus* X *cyprius,* a hybrid between *C. ladanifer,* the brown-eyed rockrose, and *C. laurifolius,* the laurel rockrose, has crinkled white petals, purple spotted at the base. *Cistus* X *purpureus,* a hybrid between *C. ladanifer* and *C. incanus,* has reddish-purple petals, yellow at the base with dark red blotches. *Cistus salviifolius* has all white petals.

FUCHSIA 'SWINGTIME' Fuchsia	Onagraceae Summer and other seasons	28°F (-2°C) Mexico, South America

Especially adapted to the milder, moister coastal areas of the subtropics, the nearly 100 species of fuchsia are all native from Mexico to southern South America except for two found in New Zealand. Many cultivars delight the eye with their crisp, jewellike blossoms in various combinations of red, pink, purple, pastels, and pure white, both single and double in form. The latter are often large, spectacular, and by far the most popular, though the singles have graceful, abundant flowers and greater resistance to sun and heat. They bloom mostly in the warmer months but in favored locations, with suitable pruning, will produce considerable winter bloom.

Fuchsia 'Swingtime,' which is illustrated, may grow to a height of 6 to 8 feet (1.8-2.4 m) with an equal spread. These growth characteristics make it adaptable for training as a formal tree, growing on a trellis, or using for a hanging basket as it is in the illustration.

Cultural requirements include well-drained soil rich in humus, partial shade, adequate watering, regular feeding with a complete fertilizer and correct pruning. Mulching is beneficial but the ground around fuchsias should not be penetrated with cultivating. Removal of spent flowers and fruits will help prolong the bloom period. Fuchsias are particularly effective for landscaping when several plants of one cultivar are massed together. They are not candidates for desert areas unless in a lath house equipped with a misting system.

Although fuchsias come from tropical areas they occur there in the cooler, moister areas at elevation and the hybrids are best in similar situations. There are a number of species in cultivation, including several of the miniatures with very small red flowers.

GREYIA SUTHERLANDII	Greyiaceae	25°F (-4°C)
Mountain bottlebrush	Spring	South Africa

The round, heart-shaped leaves and erect flower clusters of the greyias make them distinct from other plants. During most of the year, their foliage is attractive but at flowering time they are partially deciduous, revealing their stems. *Greyia sutherlandii* is by far the most attractive species. It becomes a woody shrub or small tree 10 to 15 feet (3-4.5 m) high with scarlet flowers, characterized by prominent yellow stamens, arranged in dense clusters 4 to 5 inches (10-12.5 cm) long and broad. It grows natively to 6,000 feet (1,800 m) elevation, and therefore can withstand a reasonable amount of frost. Sometimes after warm winters the old brown leaves persist and have to be removed. *Greyia radlkoferi* is similar but has velvety leaves covered with soft whitish hairs.

ALYOGYNE HUEGELII (*Hibiscus huegelii*) Blue hibiscus [*left*]	Malvaceae All seasons	27°F (-3°C) Australia

The blue hibiscus adds a delightful lilac-blue that is lacking in the color spectrum of the tropical true *Hibiscus*. It is smaller, 5 feet (1.5 m) tall, requires less water and tolerates slightly lower temperatures. The propellerlike flowers are 4 to 5 inches (10-12.5 cm) across and the leaves are rough and deeply divided. Individual flowers remain open on the plant for two or three days instead of a single day as in *Hibiscus rosa-sinensis*. It blooms at intervals throughout the year and requires occasional pruning to keep it from becoming "leggy."

HIBISCUS ROSA-SINENSIS 'AGNES GAULT' Chinese hibiscus [*right*]	Malvaceae Summer and all seasons	27°F (-3°C) Tropical Asia

The Chinese hibiscus or rose of China is the type seen in such splendor and variety in tropical countries but many cultivars are adapted to the subtropics. Efforts are continuing to develop more cold tolerant cultivars. Most are evergreen shrubs about 8 feet (2.4 m) tall. Their single or double flowers 6 or more inches (15 cm) across come in a variety of colors and appear at intervals all year.

Among some 300 species of *Hibiscus* are *H. elatus* from the West Indies, with yellow or orange-red flowers, and *H. tiliaceus*, sea hibiscus or mahoe, with yellow blossoms fading to orange-red. Both are tree forms to 20 feet (6 m) and bloom in summer.

Hibiscus arnottianus, a white-flowered species native of Hawaii, may be grown in the warmer subtropics.

Hibiscus schizopetalus, fringed hibiscus or Japanese lantern, a native of tropical East Africa, with pendent red flowers with deeply cut petals and long-extended red column of stamens, is a graceful shrub for the warmest areas. Many hybrids with *H. rosa-sinensis* may be seen in cultivation in the tropics.

Hibiscus syriacus, the rose of Sharon, is a very hardy deciduous species from east Asia, available in a variety of colors. *H. pedunculatus,* from South Africa, is 3 to 4 feet (0.9-1.2 m) high, evergreen, with small pinkish-mauve flowers like half-open trumpets, blooming during summer and fall and hardy to 25°F (-4°C).

Hibiscus may be trimmed at any time to shape the sometimes ungainly bushes and should be pruned hard at the full onset of new growth.

HYPERICUM CANARIENSE
(*H. Floribundum*)
St. John's wort

Hypericaceae
Summer

26°F (-3°C)
Canary and
Madeira Islands

Golden one and a half to 2 inch (3.7-5 cm) flowers cover this 6 foot (1.8 m) shrub during summer. Originating in the Mediterranean region it is ideally suited to similar climates as well as to warm-temperate gardens. There are many other species of shrub hypericums ranging in size to 15 feet (4.5 m), but averaging 3 to 5 feet (1-1.5 m). Most of them are evergreen and have yellow flowers.

Hypericum calycinum, H. beanii (*H. patulum henryi*), and the hybrid *H.* X *moseranum,* all attractive subshrubs 1 to 3 feet (0.3-0.9 m) high with golden yellow flowers to 3 inches (7.5 cm) across, are hardy to 10°F (-12°C). *Hypericum coris* is a pretty 1 foot (30 cm) plant useful for a ground cover but it cannot withstand severe frosts. *Hypericum* 'Rowallane' is a handsome 6 foot (1.8 m) shrub with 3 inch (7.5 cm) golden flowers.

JUSTICIA CARNEA (*Jacobinia carnea*) Brazilian plume flower	Acanthaceae Summer	28°F (-2°C) Brazil

Justicia carnea, a sub-shrub with 7 inch (17.5 cm) purplish leaves and 8 inch (20 cm) heads of densely packed flowers at the end of every stem, is a handsome, exotic addition to any shade garden where frost is not severe. In colder areas it can be grown under glass or lath and set out in late spring for color to follow camellias and azaleas.

After flowering, the stems should be cut back to two or three nodes from the ground to keep the plant bushy and to renew blooming. Cuttings root easily and with judicious pinching can be grown on to bloom as 18 to 25 inch (45-60 cm) pot or bedding plants.

Justicia rizzinii (*Jacobinia pauciflora floribunda*) makes a pretty 3 foot (0.9 m) shrub covered in winter with drooping red flowers tipped with yellow. It is tender to frost and does best in partial shade.

Several additional justicias are *J. aurea,* a shrub to 10 feet (3 m) with striking, upright, terminal yellow inflorescences to 12 inches (30 cm) long; *J. spicigera* (plants are often cultivated as *J. ghiesbreghtiana*), with scarlet or crimson flowers in loose panicles; and *J. leonardii* (*J. incana*), with rust-red flowers, a somewhat hardier, gray-leaved, 2 foot sprawling shrub, making it a showy summer ground cover or basket plant.

LEONOTIS LEONURUS	Labiatae	20°F (-7°C)
Lion's tail	Summer and Fall	South Africa

This member of the sage family has a short, woody base, sending up shoots to a height of about 6 feet (1.8 m) which in late summer and fall bear whorls of orange-yellow tubular flowers, surrounding the stem, successive tiers of flowers developing as the stem grows. After blooming, the flower-bearing stems should be pruned heavily. The flowers will keep in water for about three days if the cut stems are burned and then placed in water which has just boiled. The plants are not particular as to location, but heavy winds will break the soft flowering wood.

Leonotis nepetifolia from tropical Africa has broad, almost heart-shaped leaves and the flowers are not as neat and compact. *Leonotis laxifolia* has broad-ovate leaves distinguishing it from the lance-shaped leaves of *L. leonurus.*

LEPTOSPERUM SCOPARIUM	Myrtaceae	25°F (-4°C)
'Red Damask'	Winter-Spring	Tasmania,
Tea tree		New Zealand

There are many cultivars of *L. scoparium* available in nurseries. The illustration is that of *L. scoparium* 'Red Damask,' one of the cultivars originating in California. It grows to about 6 feet (1.8 m), is evergreen with bronze-tinted foliage and has double, almost cerise-red flowers during winter and spring.

Leptospermum scoparium 'Nanum' is a dwarf form with pink flowers; *L. scoparium* 'Keatleyi' has large 1 inch (2.5 cm) pink flowers during winter and spring, 'Snow White' has masses of pure white flowers.

Leptospermum squarrosum, an Australian species, grows to about 8 feet (2.4 m) with three-quarter inch (1.8 cm) pink apple-blossom-like flowers in late summer and fall. *Leptosperum laevigatum* is treelike to 20 feet (6 m) with single white flowers in spring and summer, shaggy bark and a bizarre twisted form. The tea trees do best in slightly acid sandy or gravelly soil and *L. laevigatum* is tolerant of coastal conditions where it is extensively used to control shifting sands.

MAHONIA LOMARIIFOLIA | Berberidaceae | 25°F (-4°C)
Chinese hollygrape | Winter-Spring | China

Probably the most decorative of the genus, *Mahonia lomariifolia* is an evergreen shrub branching close to the ground with multiple upright, thin woody stems to 12 feet (3.6 m), bearing 12 inch (30 cm) leaves divided into spiny leaflets. During winter the top of each branch is surmounted by an elongated cluster of small bright yellow flowers followed by blue berries.

Another useful species of *Mahonia* for subtropical climates is *M. bealei,* 6 to 10 feet (1.8-3 m) with large compound leaves and large plumes of fragrant yellow flowers in the spring. See page 174 for additional plants sometimes called mahonia.

MONTANOA GUATEMALENSIS	Compositae	27°F (-3°C)
Daisy tree	Winter	Guatemala

The evergreen daisy tree illustrated has successfully withstood a temperature of 27°F (-3°C) with snow. It is over 20 years old and 15 feet tall, with a treelike form but branching from the base. The peak of flowering is about Christmas, when it is covered with white, daisylike flowers with yellow centers. The striking effect of a shrubby tree covered with daisies at Christmas almost tempts one to rename it "White Christmas." This plant was introduced into the California trade as *Montanoa arborescens.*

Montanoa bipinnatifida is more shrublike, growing to about 8 feet (2.4 m) with deeply indented leaves and clusters of ball-shaped yellow and white double daisies. It blooms in late fall and early winter. *Montanoa grandiflora* is a large shrub with upright shoots bearing 3 inch (7.5 cm) single daisy flowers. It also blooms in fall and winter. After blooming, both species should be pruned back heavily, but only the dead blossoms need to be cut in the case of *M. guatemalensis.* Birds find the dried seeds a delectable feast.

NERIUM OLEANDER | Apocynaceae | 20°F (-7°C)
Oleander | Summer | Mediterranean

Native to the Mediterranean region, oleanders have for many years been popular in cultivation because of their hardiness, adaptability, summer-long colorful blossoms, and evergreen foliage. They are able to accommodate themselves to relatively dry conditions and extreme reflected heat from areas of pavement. For this reason oleander has become a popular divider plant for freeways and a screen for parking areas. They also grow and bloom profusely in areas of fairly heavy rainfall, such as Bermuda. A little extra water pays dividends in better flowering.

Oleanders are normally shrubs up to 15 feet (4.5 m), useful as specimen plants or hedges. They can be trained as trees, but continual maintenance is required to keep the new shoots pruned from the base. The white form makes a successful small tree. Different varieties make pleasing contrasts with their clusters of single or double, 2 to 3 inch (5-7.5 cm) flowers, in pink, red, salmon, yellowish, and white.

In some areas oleander is subject to attack by the oleander caterpillar as well as scale. Its good landscape qualities, however, have led it to be recommended for planting from Singapore to Cuba.

PROTEA | Proteaceae
Winter-Spring | 10°F (-12°C)
South Africa

The large family Proteaceae is found primarily in the Southern Hemisphere, chiefly in South Africa with *Protea, Leucospermum, Leucodendron,* and *Serruria,* and in Australia with *Grevillea, Hakea, Banksia, Telopea,* and *Stenocarpus.* The great diversity in the genus *Protea* led to its name for the Greek god Proteus, who changed shape at will.

In California, members of the Protea family thrive near the coast. Large plantations include many species of *Protea, Leucospermum, Leucodendron, Serruria, Paranomus,* and *Telopea* and numerous species of *Banksia.* These plants must have good drainage, full sun, and good air circulation; they do not like heavy fertilizing, excess alkalinity, or too much summer water. They tolerate cold to the mid-twenties.

Protea susannae (lower photo opposite) is said to be the easiest to grow. *Leucospermum reflexum* (upper photo opposite) has withstood a freeze of 27°F (-3°C) and 2 inches (5 cm) of snow and can bloom for about three months beginning in midwinter.

Grevillea banksii (upper right) is a 6 to 8 foot (1.8-2.4 m) evergreen shrub from Australia. Other shrub grevilleas include *G. leucopteris* (white), *G. petrophiloides* (red and green), *G. thelemanniana* (red and yellow), *G. aquifolium* (red), *G. chrysodendrum* (gold), *G.* 'Noel' (reddish), *G.* 'Constance' (orange), and *G. juniperina* 'Rosea' (rose). The grevilleas are valued as much for their handsome foliage as for their flowers.

Banksia is a remarkable genus of shrubs and trees from Australia, ranging from prostrate shrubs with stems creeping underground to trees 50 or more feet (15 m) high with brushlike clusters of flowers in various sizes, shapes and colors. The foliage varies from slender needlelike leaves to large, leathery leaves with sawtooth edges. The cut flowers last 10 to 12 days and excite admiration. *Banksia occidentalis* (upper left) and *B. media* have almost identical golden honey-colored flowers and bloom from fall to early spring. Some species flower on old wood and should receive only light pruning.

OCHNA SERRULATA | Ochnaceae | 27°F (-3°C)
Mickey Mouse plant; bird's eye bush; carnival plant | Spring-Summer | South Africa

It is difficult for the uninitiated to realize that the above pictures are of the same plant, taken about three months apart. This beautiful and versatile shrub was formerly known as *Ochna multiflora* and *O. atropurpurea.* There are about 85 species of *Ochna* including several trees such as *O. arborea* and *O. pulchra* which would be welcome additions to the landscape, but *O. serrulata* is presently the only species generally available. In warmer parts of the subtropics and tropics *O. kirkii* is cultivated. Both species grow moderately slowly and reach a height of about 8 feet (2.4 m) in full sun or partial shade. They are evergreen though the foliage may thin out in late winter. In spring new leaves are followed by masses of bright yellow flowers resembling buttercups. After about eight weeks, the petals fall and the sepals and receptacle enlarge and turn scarlet, revealing four or five shiny green fruits that turn to black as they ripen. The name Mickey Mouse plant results from the often comic face of the fruits. The color display spans a period of about five months. Birds find the ripe seed irresistible, so every year a few new ochnas spring up here and there to make fine candidates for transplanting to tubs, for gifts, hedge plants, or for topiary treatment. Both species are effective espaliered.

PITTOSPORUM NAPAULIENSE	Pittosporaceae	25°F (-4°C)
Golden fragrance	Spring	Himalayas

This shrub has sometimes been cultivated under the name *Pittosporum floribundum.* It is a vigorous grower, attaining a height of 12 feet (3.6 m) or more with a spread of at least 8 feet (2.4 m). It has large glossy evergreen leaves and in the spring puts forth masses of intensely fragrant small golden-yellow flowers in terminal clusters.

Most of the familiar pittosporums are from Australia or Asia and have white flowers. *Pittosporum tobira,* mock orange, tobira, or Japanese pittosporum, grows to 10 feet (3 m) with very thick leathery leaves and fragrant greenish-white to creamy colored flowers. Several cultivars are available, including dwarf forms and plants with variegated leaves. it is useful as a hedge, a specimen, or as a foundation planting. It is hardy, drought resistant, and may be used at the immediate coast or inland. *Pittosporum undulatum* is often grown as a shrub, but it tends to be treelike and may reach 40 feet (12 m). Other tree pittosporums, such as *P. rhombifolium* and the weeping *P. phillyraeoides,* have ornamental flowers and fruits.

PUNICA GRANATUM 'LEGRELLEI'	Punicaceae	10°F (-12°C)
Double-flowered pomegranate	Summer	Europe, Asia

The specimen illustrated is a variegated form of the double-flowered pomegranate, a large deciduous shrub or shrubby tree to 15 feet (4.5 m). The more common form of this species has large 3 inch (7.5 cm) bright orange-red flowers and blooms throughout the summer months. In the variegated form occasional flowers revert to the original orange-red, as shown in the photograph.

There is a dwarf form known as *P. granatum* 'Nana' which is only a few feet (1 m) high, has small flowers and miniature fruits. *Punica granatum* 'Flore Pleno' is a double-flowering pomegranate from southern Europe which grows to a 20 foot (6 m) small tree, has bright orange-red flowers, produces no fruit, and is hardy.

The single-flowered fruiting plant *P. granatum* appears in writings of antiquity and is believed to have been introduced into southern Europe by the Carthaginians. The dark red apple-sized fruits are divided into tiny compartments containing seeds and a red cranberry-flavored juice. When ripe, fruits pop open on the trees and are enjoyed by various birds.

Pomegranates grow best in a fertile, well-drained soil in full sun in a hot, dry climate.

PYRACANTHA COCCINEA | Rosaceae | 10°F (-12°C)
Firethorn | Spring-Winter | So. Europe-W. Asia

This hardy, large, evergreen, thorny member of the rose family, one of a number of species and cultivars of *Pyracantha,* is covered in the spring with masses of small white flowers. During the fall and early winter these are replaced by bright red fruits in large clusters which almost hide the leaves and are so brilliant and conspicuous as to thoroughly justify the name "Firethorn."

The fruits vary in color from orange to scarlet, depending upon the variety, and continue for two or three months. As they become ripe and mellow, they are regarded as delicious little apples. Let the birds enjoy these berries; they are eaten only when fully ripe and ready to fall anyway.

Pyracanthas are occasionally attacked by fire blight, causing branches, foliage, and fruit to turn brown. Since fire blight is a bacterial infection, the affected branch should be destroyed, and the pruning shears sterilized. Pyracanthas may be used as hedges, specimen plants, or effectively as espaliers.

Several other species and numerous cultivars of firethorns are available in the trade, varying in color of fruit, size of plant, and thorniness.

RHAPHIOLEPIS INDICA 'SPRINGTIME'	Rosaceae	15°F (-9°C)
India hawthorn	Spring	So. China

Although its name suggests an origin in India, this thornless, evergreen, 5 foot (1.5 m) shrub is a native of southern China. It has handsome, leathery, shiny 3 inch (7.5 cm) leaves and conspicuous masses of half-inch (1.2 cm) flowers. The flowers are followed by small purplish-black fruits. Improved cultivars with white, pink, and red flowers and a variety of habits are available.

This hardy evergreen is one of the finest shrubs for landscape use. Not only is it a good foundation plant but the flowers, especially the pinks, are excellent for cutting. It is tolerant of salt drift and shade but may be grown in full sun.

The Yeddo hawthorn, *R. umbellata,* from Japan, is somewhat taller, hardier, and white-flowered. The hybrid between the two species, *R.* X *delacourii,* has showy pink flowers.

RONDELETIA CORDATA
Heartleaf rondeletia

Rubiaceae 30°F (-1°C)
Spring-Summer Guatemala

This attractive but tender shrub is suitable only for relatively frost-free areas, especially near the coast. It belongs in a tropical family that includes *Coffea, Cinchona* (quinine tree), *Gardenia,* and *Bouvardia.* The shrub is evergreen, grows to about 7 feet (2 m) and blooms in spring and summer, with clusters of small, tubular, pink to red flowers with yellow throats.

Rondeletia amoena from Central America is slightly less showy with soft rose-pink flowers but is still attractive and can be grown in milder areas. Both species are best in good soil with regular watering and they need full sun. In cooler areas they may be used as container plants. They may be pruned to shape and to promote new growth and flowering.

LYCIANTHES RANTONNEI Solanaceae 20°F (-7°C)
(*Solanum rantonnetii*)
Paraguay nightshade Spring-Fall Argentina

This attractive evergreen semivining shrub with its masses of blue-violet, fragrant flowers affords an enticing background to the old-fashioned gate lantern shown in the photograph on page 57. One would scarcely realize that it is a relative of the weed known as deadly nightshade, or, for that matter, of the potato, tomato, or petunia. It grows to 7 feet (2 m) or more, with clusters of 1 inch (2.5 cm) flowers from spring to fall. It seems equally happy in coastal or inland regions, is drought resistant, but will tolerate average lawn conditions. It is said to provide a maximum display of flowers when given heavy feeding and generous watering, yet the specimen in the photograph was apparently growing under very dry conditions. It can be trained to a single standard, making a miniature tree.

STREPTOSOLEN JAMESONII	Solanaceae	30°F (-1°C)
Marmalade bush	Spring-Summer	Peru to Colombia

The casual observer would scarcely recognize the relationship of this gaudy, yellow to orange-red-flowered shrub to the quiet, dignified purple *Lycianthes rantonnei* depicted on page 57 and described on page 85. *Streptosolen jamesonii* is evergreen, reaches 6 feet (1.8 m), with loose, twisty branches, and blooms for several months during spring and summer. The showy flowers are in terminal clusters, the petals forming a spirally twisted tube at the base which widens out at the top. Somewhat frost-tender, this plant is best suited to milder, coastal areas. It needs sun and fast drainage. It is ideal falling over a wall or sprawling down a slope.

TECOMA STANS	Bignoniaceae	27°F (-3°C)
Yellow elder	Fall	Florida, Arizona, West Indies, Central America

This erect shrub, growing to 15 feet (4.5 m), is considered one of the finest ornamental shrubs native to the United States. It is evergreen and blooms during the fall with clusters of yellow trumpet-shaped flowers about one and a half inches (3.75 cm) long. It grows inland but is probably a little more lush and showy in coastal areas. It performs well in sites with summer heat buildup and moist soil. In colder areas, if it is frozen back it may bloom again the following year from new wood springing from the base. In warm areas it may become a small tree.

TIBOUCHINA URVILLEANA | Melastomataceae | 27°F (-3°C)
Princess flower [*left*] | Summer-Fall-Winter | Brazil

The rich royal purple of the 2 to 3 inch (5-7.5 cm) velvety flowers amply justifies the name "princess flower" for this handsome 6 to 10 foot (1.8-3 m) Brazilian shrub with its rich evergreen silky-hairy leaves. It flowers several months during summer, fall, and winter, depending upon location. It may be grown in full sun in coastal areas but in hotter, drier inland regions it is better in part shade. Good drainage and a slightly acid soil are important as are adequate moisture and fertilizer.

Tibouchina holosericea has slightly smaller purple flowers and more silky leaves. *Tibouchina mutabilis* has flowers which change from deep rose to white. *Tibouchina granulosa* is a small tree with either purple or pink flowers.

A number of members of this primarily tropical family have the potential of becoming important ornamentals yet only a few of them have been brought into cultivation.

VIBURNUM MACROCEPHALUM | Caprifoliaceae | 5°F (-15°C)
Chinese snowball [*right*] | Spring-Summer | China

The Chinese snowball, largest of the viburnums, grows to 10 feet (3 m) or more and presents a stunning sight when covered with its globose flower heads each some 5 inches (12.5 cm) in diameter, at first light chartreuse in color, maturing to pure white. The Chinese snowball has much larger clusters than the common snowball, *V. opulus,* and is not completely deciduous in winter and less plagued by aphids in summer.

Many other deciduous and evergreen viburnums are available. *Viburnum tinus,* an evergreen shrub to about 10 feet (3 m), produces tiny pinkish-white flowers in late spring. *Viburnum* X *burkwoodii,* an evergreen shrub to about 8 feet (2.4 m), has attractive clusters of fragrant white flowers.

BIXA ORELLANA | Bixaceae | 28°F (-2°C)
Annatto; achiote; lipstick tree | Summer | Tropical America

When Columbus landed in the New World he was greeted by the red Indians, red because they were said to have dyed their naturally brown bodies with annatto dye, obtained from the pulp surrounding the seeds of the annatto tree, *Bixa orellana.* Annatto dye is still used as a good coloring throughout tropical America and elsewhere. Annatto trees or shrubs are common around settlements where they are also appreciated as ornamentals with their large heart-shaped leaves and showy 2 inch (5 cm) pink to rosy flowers. The spiny yellow, scarlet, or reddish-brown fruits, opening to expose the red-coated seeds, are also ornamental. *Bixa* may grow to 20 feet (6 m) high.

IXORA COCCINEA
Flame flower; red ixora; flame of the woods; jungle flame

Rubiaceae 25°F (-4°C)
Spring-Fall East Indies

The flame flower is a fine garden shrub for the warm wet tropics and subtropics, growing to 15 feet (4.5 m), and producing broad clusters of coral red starlike flowers throughout the warm periods of the year. This ixora was introduced into England in 1690 and named for a Chinese idol. In its native land the blossoms were used in sacrifices. *Ixora chinensis,* with white, bright orange-scarlet, or red flowers, has several well-known cultivars. *Ixora williamsii* is also common in cultivation.

Ixoras may be cultivated by root suckers, cuttings, or seeds. While they will thrive in a warm wet garden they will not survive outdoors in cool wet winters.

TELOPEA SPECIOSISSIMA	Proteaceae	25°F (14°C)
Waratah	Spring	New South Wales, Australia

The spectacular waratah is the official flower of New South Wales. The dense, broadly conical inflorescence produced at the ends of the long, slender, canelike branches is 4 inches (10 cm) across with the crimson flowers subtended by brilliant pink to crimson bracts. It is valued as a cut flower.

The plants can be grown from seed preferably planted in individual containers to avoid disturbing the roots when transplanting to a well-drained, deep soil with low fertility. Pruning should occur after flowering. Waratah, like many other members of the protea family, is difficult to grow but the showy inflorescence makes it worth trying.

ACALYPHA HISPIDA | Euphorbiaceae | 30°F (-1°C)
Chenille plant; red cattail | Summer | Malay Archipelago

The chenille plant is an old greenhouse favorite, cultivated in India as long ago as 1690 and introduced into Europe in the late nineteenth century after it was discovered for the first time in the wild. It is an erect 10 to 15 foot (3-4.5 m) shrub with 12-18 inch (30-45 cm) long pendulous spikes of scarlet or pink, chenille-like flower clusters. It is a tender plant and successful outdoors only in the warmest areas.

Copper leaf, *Acalypha wilkesiana,* is grown throughout the world as an evergreen perennial shrub in the tropics and as a summer bedding plant in temperate areas for its ornamental reddish or coppery foliage. It thrives in full sun and is striking for mass foliage effects.

POLYGALA APOPETALA — Polygalaceae, Summer — 20°F (-7°C), Baja California

For dry areas *Polygala apopetala,* a shrub that may reach 15 feet (4.5 m), is recommended for cultivation. This slender stemmed shrub often is bare at the base with numerous flowers at the branch ends. Although drought tolerant, polygala will reseed in moist areas, sometimes spreading and becoming a weed.

Polygala X *dalmaisiana,* sweetpea shrub, has long been popular in European conservatories. In subtropical and tropical areas it is a satisfactory outdoor shrub, growing to about 5 feet (1.5 m) and flowering most of the year with purplish to rosy-red flowers. It is hardy to 24°F (-4.5°C) and thrives in sun or part shade.

JATROPHA INTEGERRIMA
Peregrina; spicy jatropha

Euphorbiaceae
Summer-Fall

30°F (-1°C)
Cuba

The rosy-flowered jatropha is a shapely evergreen shrub to about 5 feet (1.5 m) with clusters of rose-colored flowers to 1 inch (2.5 cm) across. The dark green leaves are often fiddle-shaped.

Throughout the tropics the genus *Jatropha* is a popular ornamental. Coral plant, *J. multifida,* native of tropical and subtropical America, is a shrub to 10 feet (3 m) with red flowers in branching clusters at the branch tips. The leaves are palmately lobed and are used medicinally along with the seeds. *Jatropha podagrica,* gout stalk, native of Central America, is a somewhat succulent plant to 5 feet (1.5 m) tall with a swollen base to the stem, giving it the common name, rounded lobed leaves, and red flowers. The seeds of all jatrophas are reputed to be poisonous.

III

FLOWERING VINES FOR YEAR-ROUND COLOR

This chapter contains some of the most colorful of the numerous vining ornamentals for enhancing fences and facades, hillsides, and retaining walls. Vines can dramatize the most effectively designed architecture and soften less pleasing structures. They can cover chain-link and drape and conceal concrete. Gardeners in subtropical and tropical climates have the advantage of being able to grow well more kinds of vining plants than those in other climatic zones.

Described in this chapter are but a few of the numerous possibilities. Purposely omitted are valuable vining materials having inconspicuous flowers such as the evergreen grapes, ivies, creeping fig, and the like. It cannot be overemphasized that chapter categories are somewhat arbitrary and many highly desirable colorful plants that are not true vines can be used successfully for a vining effect such as an espalier, but are more properly listed under shrubs and California native plants. Some examples of these are *Lycianthes rantonnei,* on pages 57 and 85; *Calliandra haematocephala,* page 62, and *Camellia sasanqua,* pages 64 and 65. Also the climbing aloes on page 162 should be noted.

Some vines (as well as some shrubs also) may be trained horizontally on the ground as ground covers, or down over cliffs or rocks as illustrated by *Bougainvillea,* pages 100, 101, 102.

When a plant of any kind is intended to clothe a wall or fence, it is imperative that structural assistance be provided. Only a very few vines will attach themselves voluntarily and even those must be spread on the new surface and encouraged with some kind of attachment. Vines climb in a variety of ways, some by means of twining stems, some by leaves or leaflets modified as tendrils, some by hooked branches. If a trellis is used it should be sturdy, adequate and attached firmly to the wall surface. Never rely on flimsy inadequate trellises. The excess portions of the plant which do not spread effectively should be removed. Plant location and pruning recommendations are given when important. Additional vines of merit are listed in Appendix III.

ACTINIDIA CHINENSIS	Actinidiaceae	20°F (-7°C)
Chinese gooseberry	Spring	China

This handsome deciduous vine of burgeoning proportions produces striking 2 inch (5 cm) white flowers that age to pale yellow. The hairy thin-skinned edible fruits are pale brown, green inside. Strong thick stems twist and twine around any structure available, making a study in rhythm and line. The deciduous leaves are some 5 inches (12.5 cm) long, dark green, conspicuously ribbed and woolly silver on the undersides. Chinese gooseberry needs some winter chilling, sun or partial shade, ample moisture, and strong support—heavy overhead structure is ideal. For fruiting it is necessary to have two plants, a male and female, or a grafted plant possessing both sexes. Prune after fruit is picked. Fruit from New Zealand is offered in markets as kiwi.

The related species *Actinidia kolomikta* is noteworthy for its ornamental variegated leaves, creamy-white, flushed pink.

ANTIGONON LEPTOPUS	Polygonaceae	15°F (-9°C)
Coral vine	Summer-Fall	Mexico

Tumbling skeins of electric pink identify coral vine in the autumn, especially in hotter regions. The abundantly massed tiny flowers cover the golden-green masses of foliage as the vine clambers unrestrainedly over fences, shrubbery, trees, and roofs, even carpeting large patches of hot bare earth as in its native Mexico. Occasionally the blossoms are softer pink, sometimes white or neon crimson. A lover of heat and strong sun, the coral vine usually performs poorly in cool gardens but is increasingly brilliant and vigorous as one travels into warmer areas. It dies to a rusty crown with frost. This means a cleanup job in the winter, though the roots survive briefly in frozen ground and produce new growth with the return of warm weather.

Coral vine is useful for covering a fence, pergola, or arbor and makes excellent cut material for flower arrangements. It should be planted in a rich, well-drained soil and given abundant water during the warm period of the year.

ARISTOLOCHIA ELEGANS	Aristolochiaceae	28°F (-2°C)
Calico flower; dutchman's pipe	Summer	Brazil

Aristolochia is a large genus of perennial plants, mostly climbers, not commonly cultivated. It usually has large heart-shaped leaves and flaring-tubed, strangely formed blossoms suggesting pipes of speckled abstract art objects. Various species occur in warm moist climates. There is one California native (*A. californica*) and at least one very hardy species (*A. durior*) native to the eastern United States, both known as dutchman's pipe. The species illustrated (*A. elegans*) is occasional in cultivation and blooms reasonably well in summer though it usually dies back in winter. It wants a warm-summer garden, ample moisture, and good soil. Its rather unbelievable 3-inch (7.5 cm) blossoms are white-veined, brownish-purple outside, purple-brown veined with yellow inside, and edged with long hairs, certainly conversation pieces. The fruits are equally interesting, resembling hanging baskets that split open when the seeds are ripe. Many of the large-flowered species would make handsome ornamental vines but they have not been brought into general cultivation.

BOUGAINVILLEA | Nyctaginaceae | 32°F (0°C)
All year | South America

No vine is more characteristic of the tropics and subtropics than bougainvillea. In frost-free areas this semideciduous to evergreen vine should be planted in far greater numbers for its spectacular color, which lasts most of the year.

Planting cultivars of different shades together can produce a scintillating color effect (see page 102). Flowers of bougainvillea are actually inconspicuous, the bright colors come from the bracts surrounding the tiny flowers. Colors range from magenta-purple through purplish pinks to crimson in *B. spectabilis* (sometimes called *B. brasiliensis*) and in the cultivars 'Rose Queen,' 'Barbara Karst,' 'Scarlett O'Hara,' 'Mrs. Butt,' 'Sanderana,' 'Texas Dawn,' and 'Crimson Jewel.' Colors range from deep bronze through orange and gold to salmon in the cultivars 'Afterglow,' 'California King,' 'Golden Glow,' and 'Mrs. Praetorius.' The selection 'Lateritia' is truly brick red. Pure white are *B.* 'Madonna,' 'Convent,' and 'Jamaica White.' The whites are very tender with 'Jamaica White' the best of the three. The dwarf 'Temple Fire' with red-purple bracts is a useful plant for the small garden. Double forms are becoming available in the trade in many areas. Innumerable cultivars have been developed in East Africa, India, Malaysia, and other subtropical and tropical areas.

Bougainvilleas are vines for the hottest possible exposures. They grow and bloom best in soil that is neither rich nor constantly watered and where the sun hits the entire plant, including its root-run, without obstruction of any kind, even ground cover or mulch. Planting against a light-reflecting wall is ideal. Most cultivars are rampant growers and require plenty of space or else frequent pruning. A few cultivars, such as 'Temple Fire,' are compact and shrubby, making good ground covers, spilling over retaining walls, or scrambling on hot rocky banks. Minimum watering after the plant is established gives maximum flowering and minimum leaf growth.

All bougainvilleas are good for growing overhead on trellises, in trees having thin foliage, or on roofs or walls, inland or near the seacoast. Also they may be trained to form hedges or clipped for formal effects. Florists have used them as pot plants. They are trained to make small street trees in Brazil where they are native and have been used in a similar way in East Africa.

Most bougainvilleas should be planted in relatively frost-free areas; however, frost-damaged plants often grow back quickly after a severe freeze. They should not be pruned or planted during cold weather. Poor soil and reduced watering late in the year may help to ripen the wood and make them more resistant to frost. For colder areas, the old familiar purple *Bougainvillea spectabilis* is best as it can stand about 10° lower temperature than other species. The bush types such as 'Temple Fire' and 'Crimson Jewel' may be grown in large containers and moved under shelter during severe freezing periods.

The upper photo on page 102 is *B.* 'Afterglow.' In the lower picture *B. spectabilis* is intertwined with an unusually red form of 'Mrs. Praetorius.' The third flowering plant in the lower picture on page 102 is *Lantana camara* (see page 149), a bright scrambling shrub. An additional bougainvillea in a rich blaze of color is shown on page 30 in front of a *Tipuana tipu* tree.

ROSA BANKSIAE 'LUTEA'	Rosaceae	15°F (-9°C)
Lady Bank's rose	Winter-Spring	China

Evergreen and usually thornless, this old Chinese species climber is timeless and useful in the garden. Best known and loved is the double light yellow form "Lutea" which may begin its long flowering season in early winter and climax in early spring with clouds of faintly fragrant small flowers in generous clusters of softest pale gold, a color especially pleasant with other spring flowers. The leaves are small, sparse, and shining, usually pestless and without mildew. Like most species roses it is an abundant grower capable of 20 feet (6 m), but it tolerates drastic pruning when blossoms are gone and this often results in another lesser wave of bloom. It likes sun or partial shade and much moisture. If pruned in winter, flowering wood will be removed. The species has fragrant white double flowers.

Another handsome climbing rose is 'Mermaid' which has single pale yellow flowers and a vigorous growth habit.

BAUHINIA PUNCTATA | Leguminosae | 28°F (-2°C)
(*B. galpinii*)
Pride of de Kaap; | Summer-Fall | Tropical and
Nasturtium bauhinia | | South Africa

This bauhinia, an excitingly showy vining shrub, is native in the eastern Transvaal, particularly in the de Kaap valley near Barberton. It is ideal for a long south-facing fence or wall, as is the blue plumbago pictured with it on the opposite page and described on page 128. It needs fast drainage, a warm root-run, and a frost-free location. Although slow to start it will attain a 20 foot (6 m) spread. Its open structure, little, light green, kidney-shaped leaves and generous clusters of two and one-half inch (6.5 cm) salmon to orange-red flowers which appear in late summer and autumn make it a most desirable garden subject. Any necessary pruning should be done immediately after blooms fade.

Bauhinia corymbosa, a native of eastern Asia, is a climbing species with pink flowers. *Bauhinia kockiana,* a native of the Singapore region, is another spectacular vine with yellow-orange flowers. Both species are recommended only for the hottest areas.

BEAUMONTIA GRANDIFLORA | Apocynaceae | 29°F (-1.6°C)
Easter lily vine | Spring-Summer | India

Beaumontia is a showy vine with white fragrant funnelform flowers 4 inches (10 cm) across in copious clusters crowning the ends of thick vining branches from midspring into summer. A heavy twiner with bold 8 inch (20 cm) rich green leaves, heavily veined, this Indian liana needs heavy support, a sunny wind-sheltered exposure and some pruning after its principal flowering. It also likes deep, rich, reasonably moist, well-drained soil and lots of room to spread even when grown as a great mounded shrub with all vining ends cut off. Easter lily vine has few pests and can be grown easily from stem cuttings that bloom sooner than plants from seed.

CLEMATIS ARMANDII	Ranunculaceae	20°F (-7°C)
Evergreen clematis	Spring	China

Evergreen clematis is a most dramatic climber when located in moist, well-drained, partially shaded spots. In early spring the dark green, prominently veined foliage is drenched with starlike white flowers of exquisite beauty. The opposite, compound leaves are distinguished by 4 to 8 inch (10-20 cm) leaflets. *C. armandii* twines rampantly into trees and cascades gracefully from roof or wall and is most successful where there is at least a hint of winter. Somewhat subject to foliage die-back, it should be pruned after bloom. *Clematis paniculata* from New Zealand with 2 to 3 inch (5-7.5 cm) white flowers is a vigorous evergreen climber.

Two of the showiest deciduous clematis hybrids grown in the subtropics are *C.* X *lawsoniana* 'Henryi' (large white) and the violet *C.* X *jackmanii.* Two North American native species, *C. crispa,* with urn-shaped bluish-purple flowers, and *C. texensis,* with similar scarlet flowers, are attractive low deciduous climbers.

COMBRETUM FRUTICOSUM | Combretaceae | 26°F (13°C)
Combretum | Summer-Fall | Tropical America

Brilliant yellow to orange flowers in groups up to 4 inches (10 cm) long by 2 inches (5 cm) wide glorify this vine from late summer through fall. The flower cluster somewhat resembles those of the bottlebrush. This woody climber is best trained on a fence or trellis, requires little care, and does well in sunny locations. *Combretum grandiflorum* with scarlet flowers and *C. paniculatum* with very showy red flowers, both native of tropical Africa, are cultivated in warm to hot wet areas such as Florida and Singapore.

CLYTOSTOMA
CALLISTEGIOIDES
(*Bignonia violacea*)
Orchid trumpet vine

Bignoniaceae | 24°F (-4.5°C)
Spring-Summer | Argentina, Brazil

Probably the most adaptable of all showy flowering climbers, the orchid trumpet vine grows easily and blooms for almost a month in sun or almost complete shade, sandy soil or clay, at the beach or far inland. It is amazingly frost-tolerant. A bit slow to start, this shiny-leaved evergreen spreads 15 to 20 feet (4.5-6 m) and grows twice as high. It is rather easily controlled by copious pruning at any season. Cascading masses of 3 inch (7.5 cm) orchid-colored trumpet blossoms usually begin in midspring and continue sporadically into autumn. It is rarely afflicted with disease or pests.

MACFADYENA UNGUIS-CATI | Bignoniaceae | 15°F (-9°C)
(*Doxantha unguis-cati*)
Cat's claw vine | Summer | West Indies and Tropical America

One of several yellow trumpet vines, the cat's claw is the hardiest of the showy self-clinging climbers. Its little clawlike appendages attach themselves to all but slickest tile to ascend at an amazing rate. Long a standby for the desert southwest, cat's claw is being used more commonly now in coastal areas where the plant is a bit slower and much easier to control. Deciduous with frost it is often nearly evergreen in mild and hot districts. In spring its new growth is palest golden green with copper tips. Each flower is 4 inches (10 cm) across, a vibrant lemon yellow.

A one-gallon-size plant installed in spring in full sun or partial shade can in six months transform an entire facade from glaring concrete to a living wall of texture and shadow pattern. This vine and lemon eucalyptus are nature's gifts to high-rise buildings and structural retainers. Once established, cat's claw is almost ineradicable and thus an efficient soil binder. Pruning may be done at any time, and it should be watered regularly.

GELSEMIUM SEMPERVIRENS | Loganiaceae | 15°F (-9°C)
Carolina jessamine | Winter-Spring | Southeastern United States to Central America

This is another most adaptable climber, seldom greedy and well suited to small gardens. It requires reasonably watered locations of sun or light shade (opposite, lower left). The small funnel-form, deep yellow flowers cascade profusely from light golden-green small foliage on willowy red-brown stems almost bamboo-like in their delicacy. Established plants in warm-winter gardens may start flowering in late fall and continue to early summer with early spring as the high point of color. The state flower of South Carolina, *Gelsemium* is completely hardy throughout most of the subtropical and warm-temperate areas. Some thinning and topping after bloom are advisable. Insects and diseases are rare.

HIBBERTIA SCANDENS | Dilleniaceae | 28°F (-2°C)
Guinea gold vine | Summer-Fall | Australia

Clear lemon-yellow single flowers, each 3 inches (7.5 cm) across, are displayed against glossy dark green leaves and twisting red-brown stems from late spring to midfall. Guinea gold (illustrated lower right), often sold as *H. volubilis,* is one of our neatest vines or ground covers. It is easily controllable, looks well the year round, and is suitable for most partially shaded moist sites. Because it is a summer bloomer one might well place hibbertia near the cool-season flowerer Carolina jessamine. Avoid locations with reflected heat or drought else thrips can be a real problem; hose off leaves often in warm dry weather. Allow 15 feet (4.5 m) minimum spread and prune top hard after flowering.

DISTICTIS 'RIVERS'	Bignoniaceae	28°F (-2°C)
Royal trumpet vine	Summer-Fall	Garden origin

Four inch (10 cm) mauve to royal purple trumpets, with yellow to orange throats, abundantly crown the branch ends of the royal trumpet vine, one of the subtropic's showiest and most vigorous warm-weather bloomers. This possible hybrid between *Distictis laxiflora* and *D. buccinatoria* (red trumpet vine) bears abundant shining dark leaves almost identical to those of the latter species. Sun-loving, intolerant of much frost, resistant to pests and disease, this vine grows easily and well in most fast-draining soils with average moisture.

Distictis laxiflora, the vanilla-scented trumpet vine from Mexico, is also a popular summer-blooming species. Its three and one-half inch (9 cm) flowers are purple, turning orchid to white, the foliage lighter green and less lustrous than that of *D.* 'Rivers.' Also the vine is less vigorous but is choice for sunny, windless sites. Prune both as the flowers wane. Both these vines are deserving of wider use throughout the subtropics and tropics.

MANDEVILLA X AMABILIS 'ALICE DU PONT' | Apocynaceae | 33°F (0.5°C)
Summer | Garden origin

This aristocratic summer-blooming tropical climber of borderline hardiness has loose racemes of clearest pink funnel-form flowers with vivid crimson throats framed by deeply veined leathery dark green leaves. In cooler areas some leaves may fall during the winter but it is essentially evergreen. It should be limited to the very mildest sites having considerable accumulated warmth. Any pruning must be done in warm weather. It must be set fairly high in fast-draining soil as cold wet root-run is lethal.

Mandevilla laxa (*M. suaveolens*), Chilean jasmine, is a 20 foot (6 m) white-flowered vine. *Mandevilla splendens* 'Profusa,' formerly listed as *Dipladenia,* is a smaller 6 foot (1.8 m) vine with rose-pink flowers. It is excellent for screening in full sun and will tolerate salt drift. Both species bloom in summer and there are no serious pests.

GLORIOSA ROTHSCHILDIANA	Liliaceae	25°F (-4°C)
Gloriosa lily	Summer-Fall	Tropical Africa

These tuberous-rooted climbing lilies are showpieces in moist or wet areas but uncommon in drier areas where they can be grown with care. They die back after flowering in summer and early fall and remain dormant all winter. They thrive only in warm-summer gardens, in full sun or partial shade, in soil that drains fast and dries out during dormancy. In very wet winters covering may be necessary. The leaves are prolonged into tendrils, the 3 inch (7.5 cm) lily flowers solitary in the leaf axils. Many blossoms may be produced by each adult tuber, and new tubers should grow from existing ones each season.

Gloriosa rothschildiana has crimson to scarlet flowers, yellow and whitish at base, the segments wavy-margined and reflexed. The flowers of *G. superba* are yellow changing to red, the segments narrow and crisp.

OXERA PULCHELLA | Verbenaceae | 30°F (-1°C)
Oxera | Variable | New Caledonia

Another distinguished tropical for gardens of the adventuresome is oxera, an evergreen climber from New Caledonia with conspicuous white 2 inch (5 cm) lightly scented flowers further characterized by long spiderlike stamens, and opposite evergreen leaves. The blossoming time is extremely variable, early spring, summer, or fall, and some years not at all. Less sensitive to cold than *Petrea* or *Mandevilla* this vining shrub is still touchy to frost so it should be placed on warmest walls away from prevailing wind, in fast-draining soil with ample humus. Only occasionally is oxera in the trade but it is well worth hunting and trying to bloom. Any pruning must be accomplished in warm weather. The red blossom shown with oxera is *Euphorbia fulgens*, a species occasional in cultivation.

HARDENBERGIA COMPTONIANA | Leguminosae | 24°F (-4.5°C)
Lilac vine | Winter-Spring | Australia

This is one of the numerous Australian pea vines, a hardy evergreen open subject for pattern work on lightly shaded fences or spilling over retaining walls. Its small violet to lilac or even pink or white flowers cascade profusely in late winter and spring. This pea vine rarely grows as rampantly as *H. violacea,* a coarser inferior subject often sold as *H. comptoniana.* It is not foolproof, however, for it often tangles, and it requires regular pruning of ungainly side branches especially after blooming. Frequent hosing off will lessen the occurrence of thrips which may disfigure foliage in rainless months.

LONICERA HILDEBRANDIANA | Caprifoliaceae | 30°F (-1°C)
Burmese honeysuckle | Summer-Fall | Burma

Cream-white blossoms, up to 7 inches (17.5 cm) long, which age yellow to gold, distinguish this giant Burmese, the largest of our many honeysuckles. The flowers occur abundantly in clusters and emit their familiar fragrance. Leaves are roundish to oval, shiningly evergreen, tender to frost. The plant grows rapidly and strong to 30 feet (9 m) in sun or light shade. Its limitations in addition to size are littering leaves that often yellow before falling, heavy twining stems often bare at the base, and a fairly invasive root system. Strong support is necessary. It is perhaps best trained horizontally on heavy wires and replaced when it becomes too bare and woody. Prune almost continually in warm weather, never in cold.

Lonicera sempervirens 'Superba,' trumpet honeysuckle, native of the eastern United States is grown for its scarlet and yellow flowers. The hybrid with *Lonicera americana, L.* X *heckrottii,* with purple and yellow fragrant flowers, is a fine garden form. *Lonicera japonica* is a vigorous grower, weedy in many areas, but it can provide a useful screen if kept under control.

PASSIFLORA | Passifloraceae | 25°F (-4°C)
Passion vine | Summer | South America

Many species of passion vines are in cultivation in the subtropics and tropics, all relatively easy to grow and reasonably hardy. *Passiflora* X *alato-caerulea* (*P. pfordtii*), shown on page 95, has foliage fairly free of predators except the caterpillar. Flowers are white-with-orchid and a crown of deep blue, purple, and white. The flowers of *P. caerulea* (illustrated opposite) are similar but smaller and the plant is more invasive and prone to caterpillars. *Passiflora edulis* is more or less deciduous, tender and often homely, grown exclusively for its fruits. *Passiflora jamesonii,* a most beautiful salmon to coral, is sensational in areas with sufficient moisture. Also tender are *P. coccinea* and *P. manicata* (lower left) with scarlet flowers with royal blue crowns. The long-tubed clear light pink blossoms of *P. mollissima* are profuse. Its invasiveness makes it desirable for cliffs and screening waste places. *Passiflora racemosa* (*P. princeps*) (lower right) is perhaps the aristocrat with pendent trusses of soft rose to coral, long-tubed blossoms. It is best where humidity is fairly high. In warmer areas *P. vitifolia* is another showy red-flowered species. All passion vines need pruning after flowering and occasional spraying for caterpillars.

SOLANDRA MAXIMA | Solanaceae | 30°F (-1°C)
Cup of gold (copa de oro) | Fall-Winter | Mexico

This luxuriant climber has large yellow coconut-scented flowers and smooth green leaves. It was long grown as *Solandra guttata, S. nitida,* or *S. hartwegii.* The flowers open light yellow and turn to gold (illustrated on page 95 with passion vine, nasturtium, and geranium). *S. longiflora,* from the West Indies, is easily distinguished by its long-tubed cream-white flowers that turn pale yellow. Both have deep brown-purple markings. True *S. guttata* has conspicuously woolly leaves.

All solandras are tender lianas, burly-stemmed, rampant, for large gardens and sturdy arbors. The blossoms are magnificent and may occur at any season, especially in winter. *S. longiflora* is usually showiest in early winter or late fall. All need considerable moisture, part to full sun, lots of room and a near frostless site not far from the ocean. All may be pruned hard in warm weather.

DISTICTIS BUCCINATORIA	Bignoniaceae	27°F (-3°C)
Red trumpet vine	Spring-Summer	Mexico

Long familiar in California gardens as *Bignonia cherere* and *Phaedranthus buccinatorius* the red trumpet vine is known for its copious clusters of large funnelform blossoms that open orange-red with yellow throats and age to rose-red before dropping. Bloom normally occurs from midspring through summer, usually most generously in locations not far from the sea. The foliage is dark green, shining, and fairly pest free. The plant is extremely robust, the top growth heavy, so it needs strong support and hard pruning after flowering. Largest flowers in greatest numbers nearly always occur on the largest leaved plants, so propagation from cuttings is desirable. This vine deserves wider use in all parts of the subtropics and tropics.

PODRANEA RICASOLIANA	Bignoniaceae	32°F (0°C)
Pandorea; Port St. Johns creeper	Summer	South Africa

Another trumpet vine, this one pink-striped crimson, is occasionally sold as *Bignonia mackenii.* Lacy foliage of considerable refinement drops with a hint of frost. Carefully pruned *P. ricasoliana* can be stunning in summer spilling from an arbor or off hot rocks. It is widely used in south Florida, Cuba, and Costa Rica.

Podranea brycei, Zimbabwe creeper, a summer bloomer in its native Africa is a winter bloomer in California where it is rare in cultivation. It has large clusters of pale pink fragrant flowers marbled red with yellow throats. Leaves are deciduous and the plant tender. Both species are best in a sheltered site in full sun with rich soil and heavy mulch of fertilizer.

PYROSTEGIA VENUSTA | Bignoniaceae | 28°F (-2°C)
Flame vine | Summer-Fall-Winter | Brazil-Paraguay

Strictly for hottest gardens with reflected heat, this climber, also sold as *P. ignea* and *Bignonia venusta,* reaches its climax about Christmas in north temperate areas. Especially happy on tile roofs whose color is convivial with the sheets of flame-orange narrow-tubed flowers that obliterate the usually sparse foliage, flame vine drips brilliantly from eaves and can curtain concrete walls that face the sun. Hopeless in shade and dubious in most of the fog belt *Pyrostegia* is sensational when happily sited. It is one of the most frequently seen vines throughout the tropics and subtropics. Any pruning is best accomplished in warmest weather.

SENECIO ANGULATUS | Compositae | 25°F (-4°C)
| Fall, Winter | South Africa
SENECIO CONFUSUS | | 32°F (0°C)
| Spring to Fall | Mexico

Two of the numerous climbing senecios are *S. angulatus* and *S. confusus* (opposite, lower left). *S. confusus* blooms red-orange with thickish bright green leaves. Best in coastal gardens with fairly moist air and no frost, it also appreciates its roots in some shade, its top in half a day's sun. Keeping old flower heads cut is a chore but worth the effort. Cut the entire vine back hard when through blooming; it is slightly dormant in winter.

The somewhat succulent *S. angulatus* (lower right) is yellow, with shining evergreen leaves. It is shown with *Solanum dulcamaroides,* violet with yellow centers, and with red *Cotoneaster* berries.

STEPHANOTIS FLORIBUNDA | Asclepiadaceae | 32°F (0°C)
Madagascar jasmine | Late Summer | Malagasy

The enduring elegance of form, texture, and fragrance in *Stephanotis* is rivaled in few ornamentals. Each 2 inch (5 cm) blossom is composed of five waxen petals that flare from an ivory tube. The flowers cluster loosely in groups of five to nine, startling white against shining dark green leaves (see upper left), in late summer.

Successful placement means shaded roots, head in at least half sun, moist rich soil that drains fast. Sandy loam with redwood shavings and leaf-mold provides an ideal growing medium. On a wind-sheltered wall this plant's well-groomed appearance is a delight at all seasons. But it can be damaged by frost, and water-logged ground will shorten its naturally long life. Light feeding and any pruning should be done in warm weather.

STIGMAPHYLLON LITTORALE | Malpighiaceae | 26°F (-3°C)
Orchid vine; golden vine | Spring-Fall | Brazil

The orchid vine, *Stigmaphyllon littorale,* shown upper right is more rugged than its slender twining relative from the West Indies, *S. ciliatum.* Its larger, darker green leaves are somewhat hairy and coarse. The lemon-yellow, green-tinged, conspicuously clustered flowers are produced recurrently over many months. Excellent on hot banks where the robust tuberous roots can check erosion, this species will cover chain-link fences. *Stigmaphyllon ciliatum* is a lighter weight vine blooming profusely and intermittently from spring to fall, with exquisite small clear yellow flowers suggesting little orchids in form. It should be planted in well-aerated soil with ample humus. Prune in warm season.

TRACHELOSPERMUM JASMINOIDES | Apocynaceae | 15°F (-9°C)
Star jasmine | Spring-Summer | China

Star jasmine (shown in the lower picture opposite) is one of the commonest and most versatile among evergreen twiners. It serves either as a vine, a low shrub, or a ground cover, and is hardy. Its white lacy starlike flowers, enchantingly fragrant, are profuse in spring and summer, and contrast handsomely with dark green polished 2 inch (5 cm) leaves. By nature a rambler, the plant may be trained to curtain walls, spill from plant boxes, climb posts and trees, and cover fences and eaves. A bit slow to start, its life span is long. Shear unwanted tendrils at any season.

Trachelosperum asiaticum, the yellow star jasmine from Japan and Korea, is a rarer species with slightly broader leaves. It is less vigorous and grows more slowly.

The genus *Jasminum,* in the olive family, contains a number of the true Jasmines. *Jasminum angulare* from South Africa (erroneously known in California as *J. azoricum*) has white flowers from summer to winter, likes a sunny location, and is frost tender at 30°F (-1°C). *Jasminum humile* 'Revolutum,' the yellow bush jasmine from tropical Asia, will withstand cold to 10°F (-12°C), blooms in summer, likes a sunny location, and achieves a height of 10 feet (3 m) when used as a shrub. The primrose jasmine from China, *J. mesnyi* (*J. primulinum*), is also hardy, has yellow flowers from late fall to early spring, likes a partially shaded location, and will extend to 15 feet (4.5 m). The Spanish jasmine (actually from Iran), *J. grandiflorum,* grows to 15 feet (4.5 m), has white flowers in the summer, is hardy to 28°F (-2°C) and also likes partial shade. The Chinese jasmine, *J. polyanthum,* grows to 20 feet (6 m), has fragrant white blossoms, rose-colored outside, blooms from midwinter through summer, likes a partially shaded, moist location, and is somewhat more tender.

SOLANUM WENDLANDII	Solanaceae	25°F (-4°C)
Costa Rican nightshade	Summer	Costa Rica

In hot protected sites *Solanum wendlandii* produces a summer spectacle of lavender-blue two and one-half inch (6.5 cm) blossoms in large branched clusters, yellow-eyed and conspicuously handsome. Though best in full sun in protected coastal areas, it may flower inland when grown in part shade. It should be pruned immediately after flowering. Deciduous during cold weather.

Solanum jasminoides, the white sweet-potato vine, is hardy to 20°F (-7°C), has bluish-white, small, star-shaped, yellow-centered flowers that occur intermittently all year, sometimes sparingly, sometimes in cascades. It is best in slight shade with moisture and occasional heavy pruning at any season.

Solanum dulcamaroides (*S. macrantherum*) is shown on page 121 growing with yellow *Senecio angulatus* and *Cotoneaster.* It is an exquisite climber with long-stemmed emerald to light green foliage, graceful twining branches and yellow-centered, vividly violet flowers, conspicuously clustered. Little red berrylike fruits follow the bloom which may occur at any season. A wind-sheltered sunny location is best.

WISTERIA SINENSIS | Leguminosae | 10°F (-12°C)
Chinese wistaria | Spring | China

The wistarias, appreciated for their striking pendulous clusters of pea-like blossoms in spring, are large, deciduous vines for sun or partial shade. Flower buds are encouraged with periodic pruning after flowering which will also restrain these vigorous vines to moderate size. They may even be trained as shrubs or small trees. Root pruning may be necessary to shock plants into setting flower buds.

The Chinese wistaria, illustrated opposite, has fat clusters of lilac or white blossoms. The Japanese wistaria, *W. floribunda,* produces hanging flower clusters one and a half feet (45 cm) long, its cultivar 'Longissima' skeins to 4 feet (1.2 m). The silky wistaria, *W. venusta,* has white flowers, and its cultivar 'Violacea' fragrant purple-blue flowers in short heavy clusters.

THUNBERGIA GREGORII | Acanthaceae | 30°F (-1°C)
All year | Africa

Thunbergia gregorii, upper picture, and *T. gibsonii* are lightweight climbers or crawlers grown for their curtains of brilliant orange blooms in full sun, mostly in summer and fall. *Thunbergia gregorii* with richer evergreen, somewhat hairy foliage blooms heavily in hot weather and often throughout the year. *T. mysorensis* from India produces yellow and reddish flowers opening in succession along the pendent threadlike flower stems. The black-eyed clockvine, *T. alata,* is known for its variable flower colors, white to orange or even maroon with shiny black centers.

Thunbergia grandiflora, the blue sky vine or Bengal trumpet vine, from India (lower right) has ropes and skeins of light blue, two-lipped funnels dangling from pale green leaf masses in autumn. This plant should be pruned back hard after blooming. Any frost will blacken the 6 inch (15 cm) leaves but recovery is rapid in the spring. Creditable flowering requires months of accumulated heat making this a particularly good vine for the tropics. The leaves are subject to yellowing. This is no plant for limited spaces but grown on a large and sturdy arbor it can be dramatic. A white-flowered variety is also available. All thunbergias should be placed in sunny locations in deep porous soil away from wind, and pruned only in warm weather. All are pest-free.

TECOMARIA CAPENSIS | Bignoniaceae | 25°F (-4°C)
Cape honeysuckle | Summer-Fall | South Africa

This is a rapid-growing evergreen vining shrub (lower left) with lustrous dark green leaves handsome at all seasons. It flowers brilliantly in sun through summer and fall with upright clusters of red-orange tubular funnels. *Tecomaria* grows well in dry or wet soils and sand or heaviest clay and functions in the garden as a bush, vine, or ground cover. Heavy pruning in early winter after most of the bloom is past should keep it from being invasive, but it can be cut again at any time. From seashore to desert it is completely successful and usually free from pests and disease. It will even thrive in shade though it seldom blooms there. There is also a yellow-flowered form, less vigorous with fewer smaller blossoms. Selections are available that are dwarf, have bronzy foliage, and deeper red flowers.

PLUMBAGO AURICULATA (*P. capensis*)	Plumbaginaceae	25°F (-4°C)
Blue Cape plumbago; embeleso	Summer-Fall	South Africa

The clear light blue masses of phlox-like flowers of plumbago may be seen in many parts of the subtropics as well as in the tropics. The cultivar 'Alba' has pure white flowers. From spring to late fall this vining shrub romps freely, sometimes aggressively in sun or light shade. Seldom at its best in small gardens for it is not by nature neat, it can be confined to island beds, pots, or trimmed as a hedge or shrub. Plumbago is a large-space ornamental that benefits from occasional heavy shearing, especially in winter when somewhat dormant. Native of South Africa, it has long been naturalized in warm dry regions. It is shown with *Bauhinia* on page 105. It is pictured growing over a cement block wall with *Spartium junceum* (Spanish broom) and red geranium.

ALLAMANDA CATHARTICA | Apocynaceae | 25°F (-4°C)
Common allamanda; golden trumpet | Summer | Tropical America

One of the most common ornamental vines of the tropics, this evergreen, shrubby climber has glossy green foliage and yellow bell-shaped flowers to 5 inches (12.5 cm) across. In hotter areas allamanda is in flower intermittently throughout the year; in cooler areas it flowers only with heat buildup in the summer and autumn months and may be completely deciduous. A number of cultivars are available, 'Grandiflora,' a compact dwarf with lemon-yellow flowers; 'Hendersonii' with flowers tinged brown in bud; 'Schottii' with flowers having a dark-striped throat; and 'Williamsii' having flowers with a reddish-brown throat. Allamanda is rampant in the tropics, flowering best in full sun, and useful as a covering for pergolas, archways, and pillars. It needs heat and is best in the warm wet tropics where a temperature below 50°F (10°C) is rare; however, it will return quickly after a light freeze.

Allamanda violacea, purple allamanda from South America, has dusty-rose to reddish-purple flowers to about 2 inches (5 cm) across. It is a scandent shrub, not as rampant as *A. cathartica,* and can be easily pruned to remain as a shrub. It is suited to dry conditions.

CONGEA TOMENTOSA	Verbenaceae	25°F (-4°C)
Shower orchid	Summer	Burma, Thailand

This showy climbing shrub has leaves to 8 inches (20 cm) long, tomentose beneath and white flowers with persistent white to pinkish-lilac bracts. It has been cultivated in areas with a Mediterranean climate. The related *Congea velutina* from Malaya is similar. The panicles of the inflorescence have clusters of flowers, each subtended by woolly 3 to 4 inch (7.5-10 cm) long lilac-colored bracts and each flower has a woolly white calyx. The delicate, almost smokelike quality of this vine makes it a good background for more brilliant flowers.

PETREA VOLUBILIS	Verbenaceae	30°F (-1°C)
Queen's wreath; sandpaper vine	Spring, Summer	Tropical America

Sandpaper vine is a high climbing woody vine to 35 feet (10.5 m) or more. The common name describes the roughened leaves. The bluish-lavender flowers are borne in clusters 3-12 inches (7.5-30 cm) long. The tubular corolla is subtended by a usually blue-lavender calyx that enlarges and persists long after the corolla has fallen. The vine is best in full sun and particularly where it can get reflected heat. It is easily propagated by cuttings.

A related species, *P. arborea,* with blue flower clusters 2 to 6 inches (5-15 cm) long, can be grown as a shrub or low tree although it is also somewhat vinelike. White flowered forms are occasional.

IV

FLOWERING PLANTS FOR COLOR ON THE GROUND

The plants in this chapter are those that can be used as ground covers, defined here as plants that can add color to the landscape near ground level, especially when planted in mass. Effective use of color can be achieved with annuals, biennials, perennials, many types of succulents, vines, and low shrubs. Many different types of plants can be massed to make excellent ground covers.

Ground covers have several functional values in addition to the color emphasized here. They serve as a method of weed control; add texture in the composition of landscape design; soften the effect of cliffs or other rugged abutments; control erosion, particularly on slopes; can provide a fire-retardant border; and serve as substitute for lawns. When low maintenance is a consideration, low, spreading ground covers are the answer. It should be emphasized that additional ground covers are to be found in the chapters on shrubs, vines, and California native plants.

Included here are some plants, such as the heliconias and gingers, that are normally not considered as ground covers, yet at the base of slopes planted in mass, they serve that purpose and provide a show of colorful flowers.

CATHARANTHUS ROSEUS — Apocynaceae — 27°F (-3°C)
(*Vinca rosea*)
Madagascar periwinkle — Spring-Fall — Tropics

This perennial shown opposite is treated as a low bushy annual bedding plant in colder areas. In milder zones where it is handled as a perennial it may not be attractive in the winter but will be a mass of bloom, like the illustration, through the spring, summer, and fall. In the tropics it may be in flower throughout the year. Ordinarily the foliage is a handsome glossy green and the flowers are available in a variety of color patterns of white and rose. This plant loves heat and is best in sun or light shade. It is often used as a bedding plant in parks.

Catharanthus is additionally interesting as the source of a drug that is being used in the treatment of certain types of leukemia.

DIMORPHOTHECA and OSTEOSPERMUM	Compositae	20°F (-7°C)
African daisies	All year	South Africa

The closely related dimorphothecas and osteospermums are popular garden plants and many of their cultivars can be used effectively as ground covers. The cultivar illustrated and known as *Dimorphotheca* 'Buttersweet' may be a hybrid between *Dimorphotheca* and *Osteospermum.* It has shrubby growth to two and a half feet (75 cm) and benefits from occasional hard pruning.

Osteospermum fruticosum, the trailing African daisy, shown upper right with bronze-leafed New Zealand Flax, *Phormium tenax,* is characterized by quick trailing growth and purplish flowers that appear in winter and spring. The typical wild form is said to have white flowers. The rather fleshy leaves are light green and plants reach one and a half (45 cm) feet in height. 'Snow White' is a selection with more upright growth and nearly white flowers.

In 1967 the Los Angeles State and County Arboretum introduced another selection called *Osteospermum fruticosum* 'Burgundy Mound,' shown upper left. This forms cushion-shaped plants with wine-colored ray flowers. All of these require full sun and will tolerate only light traffic.

Dimorphotheca pluviolis (*D. annua*), Cape marigold from South Africa, is a 16 inch (40 cm) annual plant with white and violet flowers during the winter. *D. sinuata,* another South African annual daisy, grows to only 12 inches (30 cm) and produces orange-yellow flowers in winter and spring.

POTENTILLA CRANTZII
(*Potentilla verna*)
Spring cinquefoil

Rosaceae 0°F (-18°C)
Spring-Summer Europe

The spring cinquefoil is valuable for its adaptability to many climates. It is a low creeper with clusters of bright yellow flowers in spring and summer, making a dense ground cover that spreads rapidly by runners in sunny locations having well-drained soil. It is an excellent cover for dry slopes. *Potentilla cinerea,* alpine cinquefoil, with pale yellow flowers and *P. tridentata,* the wineleaf cinquefoil, with white flowers are also hardy.

AJUGA REPTANS	Labiatae	15°F (-9°C)
Bugle weed	Spring-Summer	Europe

Ajuga is one of the most satisfactory ground covers for shade, but will endure some sun in coastal areas if moisture is adequate. It will withstand occasional light foot traffic and will spread and cover well from a thin planting. A number of interesting cultivars are available. 'Purpurea' has bronzy purple leaves that provide a strong note of color for the landscape designer. 'Variegata' has white-edged leaves. A giant-leaved form, *A. reptans* 'Crispa,' is also available. This plant is not sensitive to frost. Attractive blue flowers appear on the spikes in spring and early summer.

CONVOLVULUS SABATIUS (*C. mauritanicus*) Ground morning glory	Convolvulaceae Spring-Summer	15°F (-9°C) North Africa

Blooming for a long period during the spring and extending into the warmer months of the year, this attractive low-growing perennial plant has hairy gray-green leaves and bluish-lavender flowers. It requires a well-drained soil and the plants should be thinned out occasionally to encourage new growth. It is useful in both solid masses and mixed plantings and is particularly good for hot, dry situations. It is also usable as a hanging basket plant. This convolvulus is grown in many areas of the subtropics.

In the photograph the low marigolds provide a color contrast to the blue of *Convolvulus.*

Convolvulus cneorum, a 4-foot shrub with white flowers, blooms from spring into early fall in full sun with fast drainage.

POLYGONUM CAPITATUM | Polygonaceae | 28°F (-2°C)
All year | Japan

Characterized by abundant small pink flower heads and a bronzy foliage, this attractive knotweed (in foreground above) has been popular in the mild climate areas because it requires little attention and is colorful throughout the year. Somewhat weedy, it is best in isolated areas where it can be confined and where there will be no foot traffic. Often it will reseed itself naturally after a freeze. It is an attractive plant when draped over a wall.

AGAPANTHUS AFRICANUS | Amaryllidaceae | 23°F (-5°C)
Lily-of-the-Nile | Summer | South Africa

The common name of this blue-flowering plant, shown in the left of the above photo, is misleading since it does not come from the northern part of Africa. It does, however, prefer abundant watering when in flower, as the name might suggest. A wide selection of agapanthus species and cultivars is available, varying in size, habit, and color of the flower. There are both deciduous and evergreen, dwarf and tall types, and colors range from white to blue and purple. These plants produce cut flowers for spectacular flower arrangements. They are adaptable to many locations and particularly effective as container plants in the patio.

The large plant on the right is *Doryanthes palmeri,* a native of Australia, which puts forth a tall spike of red flowers resembling the torch ginger.

ARCTOTHECA CALENDULA
Cape gold

Compositae 28°F (-2°C)
Spring-Winter South Africa

Extremely vigorous, *Arctotheca* quickly spreads by rooting stolons. With water and if unchecked a small plant can cover as much as 200 square feet (18 sq. m) in a year or two. It should be introduced with caution to areas having frost-free winters and rainy summers. Fortunately only a single self-sterile clone is available and it therefore cannot spread by seeds.

The mat of silvery green leaves varies in thickness according to availability of water. In sunny places large bright yellow flowers are produced profusely most of the year. Coarse and not as refined as many ground covers, Cape gold is best used to cover large slopes where less robust plants need not compete with it.

PRIMULA X POLYANTHA
English primrose

Primulaceae 20°F (-7°C)
Winter-Spring Europe

The English primroses are notable examples of the progress made by plant breeders. Many unusual types with a wide variety of flower colors, ranging through the spectrum, are available in the seed trade. They are one of the best bedding plants for a mass display of color in shaded areas. *Primula* plants may remain in the ground several years before it is necessary to divide the clumps and start them over. They make excellent cut flowers or potted plants and are a favorite the world over in window boxes. Moist well-mulched slightly acid soil is required, and plants must be protected against slugs and snails.

FRAGARIA CHILOENSIS | Rosaceae | 10°F (-12°C)
Wild strawberry; sand strawberry | Spring-Summer | Pacific Coast, North and South America

This wild strawberry of new world origin has leaves of a deep shiny green. It has been widely used as a lawn substitute where only light foot traffic may be expected. In the fall the leaves take on attractive reddish tones (upper left).

Occasional mowing, especially early in the season, will improve the quality of the cover. Although this strawberry is somewhat drought resistant, it is best with watering and fertilizing. However, overwatering may cause attacks of disease and good drainage is important. Iron sulfate is used to control chlorosis (yellowing of the foliage). Occasional replanting is desirable as the plants get older.

Fruiting is ordinarily sparse but the introduction from Rancho Santa Ana Botanic Garden, 'Hybrid No. 25,' is similar to the species and produces some usable strawberries. This cultivar should be kept thinned to favor production of fruit.

CERASTIUM TOMENTOSUM | Caryophyllaceae | 0°F (-18°C)
Snow-in-summer | Spring-Summer | Europe

This low creeping hardy perennial is frequently used in rock gardens and borders in the colder parts of the country. Abundant white flowers appear in spring and summer, their color enhanced by grayish, hairy foliage. The plant prefers sun, spreads rapidly in favorable situations, and some trimming is needed when it is grown for several years. It is valuable for cold mountain areas. (Shown upper right.)

The effectiveness of *Cerastium* may be enhanced by contrasting materials such as a clumping grass, the blue fescue, *Festuca ovina glauca,* shown with it. This hardy plant requires well-drained soil and some sun for best results.

CENTRANTHUS RUBER | Valerianaceae | 15°F (-9°C)
Jupiter's beard; valerian | Spring-Fall | Mediterranean area

Here is a horticultural Cinderella that can become a glamorous fairy princess when used in the right setting! It is slightly weedy but not objectionably invasive and it will give a lavish display of color over long periods with very little water and care. It can be used boldly in large masses, for this is an ideal low-maintenance plant. It is benefited by occasional thinning but will persist year after year in the same location. This is a good plant for steep, dry slopes and unfavorable soils. The flowers range from various reds and pinks to a pure white and are produced in profusion over a long period of time.

The effective but informal use of this plant in front of a 100-year-old building is illustrated in the lower photograph on the opposite page.

ECHIUM FASTUOSUM	Boraginaceae	22°F (-5.5°C)
Pride of Madeira	Spring	Canary Islands

Bold plants which produce spectacular foliage or flower effects can be used as ground covers, especially on slopes, as the photo demonstrates. This unusual plant has irregular spreading stems and large spikes of 6 feet (1.8 m) of purple or dark blue flowers that are produced in late spring above the foliage. The plant requires little care or watering and is especially good near the seacoast, but it will thrive in other areas. Structural forms can be developed by pruning. It can be used effectively against a hedge or wall. Some closely related species with different flower colors are available. Individual plants may not be long-lived, but they reproduce readily from self-sown seedlings.

FELICIA AMELLOIDES 'SANTA ANITA' | Compositae | 20°F (-7°C)
Blue daisy; agathea | Spring-Fall | Africa

This cultivar of the old garden favorite, the blue African daisy, was developed at the Los Angeles State and County Arboretum by increasing the chromosomes of the plant through treatment with colchicine. This increased the flower size and substance considerably. It is worthy both as a ground cover in the garden and as a cut flower (lower left). The cultivar 'San Gabriel' is more shrubby. Other cultivars include 'Jolly,' 'Dwarf,' and 'Elisabeth Marshall.'

FELICIA FRUTICOSA | Compositae | 26°F (-3°C)
Aster bush | Spring | South Africa

The bush aster (earlier placed in *Aster* or *Diplopappus*) is a somewhat woody plant with a height and spread of several feet (1 m) and rather fine but dense foliage. In the spring it is spectacular with the entire shrub covered with purplish-blue flowers. Some thinning out of the old growth is advisable to keep the plant attractive throughout the year. The plant likes sun and requires little water. (Shown lower right.)

EUPHORBIA RIGIDA
(*E. biglandulosa*)

Euphorbiaceae	20°F (-7°C)
Winter-Spring	Mediterranean area

This low growing herbaceous perennial produces attractive chartreuse-colored flower clusters at the top of stems with gray-green leaves in late winter. It is a good plant for those who prize unusual effects or who want a variety of materials for floral arrangements. After flowering, new stems appear and the old stems should be removed to maintain the best appearance at all times of the year. Full sun and good drainage are best.

HEUCHERA
Alum root; coral bells

Saxifragaceae	10°F (-12°C)
Summer	North America

Heuchera sanguinea and its cultivars are well-known garden favorites in a genus that includes species from various parts of North America. A few of these species are California natives, and some hybrid forms of these produced at Rancho Santa Ana Botanic Garden are particularly showy.

These plants are good in shade and prefer moist soils. They form low mounds of attractive leaves and bear beautiful panicles of red, pink, or white flowers in the summer. They are unusually graceful plants in the garden and also provide colorful cut flowers. Heucheras are easily propagated by divisions.

Heucheras are shown here with *Artemisia pycnocephala* and *Nemophila maculata.*

GAZANIA 'COPPER KING'	Compositae	27°F (-3°C)
Gazania	Spring-Fall	South Africa

Gazanias provide a colorful ground cover for areas free of foot traffic. They are best suited to warm sunny situations and do not endure prolonged freezing. Hybridizers have greatly extended the color range available to include from white and yellow to pink, orange, and brownish-red, of which the popular variety 'Copper King' is an example.

Most of the older cultivars were derived from species that have a clumping habit. These spread slowly and need to be planted about a foot (30 cm) apart to make a solid effect. Trailing types from *Gazania uniflora* have been introduced. These grow rapidly and produce long stolons that permit a few plants to spread quickly over large areas. This species has somewhat woolly foliage.

KNIPHOFIA UVARIA	Liliaceae	20°F (17°C)
Torch lily; red-hot poker	Spring-Summer	South Africa

Kniphofia, formerly called *Tritoma,* somewhat resemble the aloes in flower form. There are about 70 species of these herbaceous perennial plants, native to eastern and South Africa. Hybridization has produced a striking array of colors including white, chartreuse, yellow, gold, bright orange, and flaming red. They are superb as cut flowers for their long-lasting quality, also frequent cutting stimulates flower production. The foliage is an attractive clump of long arching lily-like leaves varying in size.

Small plants with dainty spikes 1 foot (30 cm) high make excellent borders. Others produce spikes as high as 6 feet (1.8 m). Modern hybrids are robust and do not require staking. By planting a variety of cultivars, the flowering season may be extended from early spring through summer with individual plants having as many as 10 to 15 flower spikes.

Full sun and rich well-drained soil with moderate watering is all that is required. They respond well to standard commercial fertilizers and are remarkably free of pests. They can be subdivided annually in late fall or early spring. Seeds are easily germinated and the young plants will bloom the following year. For the amateur hybridizer, *Kniphofia* offers an exciting challenge as the various species cross readily and plants grown from the seed thus produced vary widely.

STRELITZIA REGINAE	Strelitziaceae	24°F (-4.4°C)
Bird of paradise	Fall-Spring	South Africa

Unique in color and structure, this striking flower is the official city flower of Los Angeles. It is highly recommended for areas without frost. The plants vary considerably from seed and are best purchased when in bloom for color selection. They propagate readily by division and are easy to grow, preferably in full sun with a high level of fertilization, and can be adapted to container growing. The flowers are long-lasting when cut. Plants bloom about six months starting in midfall.

A treelike species is *Strelitzia nicolai* with bluish and white flower petals.

LANTANA MONTEVIDENSIS	Verbenaceae	26°F (-3°C)
Trailing lantana	All year	South America

This species of lantana has long trailing stems with small leaves and bears clusters of lavender-purple flowers in profusion throughout the entire year. It is considered tender but is sometimes seen in interior valleys in sheltered locations. It withstands much drought and neglect and is one of the most satisfactory and ornamental plants for slopes in areas with favorable climate. It is best in full sun and is attractive when draped over a wall. An undesirable wild seedling form should be avoided as it is an invasive weed. For a spectacular bloom over a whole year it is difficult to surpass the bush or trailing lantanas.

Flowering aloes and yellow *Cassia artemisioides* are seen in the background.

LANTANA CAMARA
Bush lantana

Verbenaceae
All Year

26°F (-3°C)
American tropics

The bush lantanas, particularly the dwarf forms, function well in many types of plantings. They can be treated as annual bedding plants in cold areas. Many named cultivars of this species and of hybrids with *Lantana montevidensis* offer a size range from 1 to 6 feet (0.3-1.8 m) and flower colors in yellow, pink, white, orange, and red.

Lantanas are tolerant of heat and drought but may occasionally be attacked by insects. They are notable for producing abundant flowers over a long season. These plants adapt so well that they have run wild in parts of the tropics, particularly Hawaii, where they are serious pests. For dependable color over long periods few other plants can match the many lantana varieties.

LIMONIUM PEREZII
Sea lavender; statice

Plumbaginaceae
Spring-Fall

28°F (-2°C)
Canary Islands

This attractive herbaceous perennial from the Canary Islands is so well adapted to the climate of coastal areas that it has become naturalized in many places on rocky cliffs near the seashore. The long-lasting purplish-blue flower clusters are produced over a very long period. The plant does not stand much frost, but where temperatures permit, it deserves far wider use than it has had.

The photo shows sea lavender with *Chrysanthemum coronarium,* an annual frequently naturalized along the coast where it forms large masses of color.

PUYA SPATHACEA | Bromeliaceae | 27°F (-3°C)
Early Summer | Argentina

Shown in the picture above is one of the outdoor bromeliads grown in milder areas of the subtropics with spectacular effect. This relative of the pineapple has deep blue-green flowers but its rose-pink bracts and branching reddish flower stalks 2 to 3 feet (60-90 cm) high are more showy. It needs well-drained soil and must not be overwatered. *Puya benteroniana* (in trade as *P. alpestris*) is taller with larger flowers having an unforgettable metallic blue-green color reminiscent of peacock feathers. *Puya chilensis* with chartreuse flowers is even more spectacular.

PELARGONIUM	Geraniaceae	27°F (-3°C)
Geranium	Spring-Fall	South Africa

Ivy geranium is used widely in mild coastal areas for spectacular effect. A species with trailing stems, somewhat fleshy succulent leaves, numerous flower types and colors has special appeal to collectors. White, pink, lavender, and rose are colors most commonly seen. They require some attention including periodic replanting to keep them in good condition. On page 153 they are seen with yellow *Chrysanthemum coronarium.*

Illustrated above is *Pelargonium* X *domesticum* (Martha Washington geranium) which has a great variety of flower color and form with dark blotches on the two upper petals. Plants are erect to 3 feet (90 cm) or more and somewhat spreading. The showy flowers may be more than 2 inches (5 cm) across in loose clusters.

Pelargonium X *hortorum* shown above and on page 128 with *Plumbago auriculata* is the most common of the garden geraniums. It has varicolored leaves with flowers smaller but in fuller clusters than in Martha Washington types. It is used widely for outdoor massed color effects.

SUCCULENTS FOR COLOR

The word *succulent* is a somewhat arbitrary term used for a very large number of plants that have in common thick fleshy leaves or stems making it possible for them to withstand an arid environment. All cacti are succulents but not all succulents are cacti. Because their water and care requirements are minimal while their landscaping effect can be maximal, they have become an important part of the landscape in the more arid areas of the tropics and subtropics. Most prefer a mild climate but some can endure cold. And some thrive in the wetter areas if good drainage is provided.

Most succulents are started easily by cuttings placed in the soil during periods of rainfall or irrigation. Highway plantings are started in this manner in areas near the coast. Sometimes the rosette types such as *Echeveria* are planted close together to produce a sheet of almost solid color from their foliage alone.

The succulents require little maintenance or pest control through dead stems and foliage should be cleared away. Succulents of appropriate types often are an ideal solution to landscape design when color at ground level is the objective. All plants described and pictured on pages 155 to 163 are succulents. They are propagated easily by cuttings rooted in sand.

LAMPRANTHUS SPECTABILIS | Aizoaceae | 25°F (-4°C)
Trailing ice plant | Spring | South Africa

Among the most showy of succulent ground covers are species of *Lampranthus*, such as *L. spectabilis* (page 155 and above), with purple flowers in great profusion, and *L. aurantiacus*, with large orange or yellow flowers. When out of bloom these species have great practical value as ground covers though with less character than when glowing with flowers. Most are best restarted every three or four years from cuttings that root easily in damp sand or even directly in the ground.

Malephora purpureo-crocea and *Delosperma* 'Alba' are recommended for highway plantings because their inconspicuous flowers do not distract motorists. In gardens other species are better for their splashes of brilliant color as illustrated on page 155 and above. *Drosanthemum speciosum* (page 155) is a small shrub with striking flowers of burnt orange with green centers. In contrast, *D. floribundum* forms thin purple-flowered sheets which hug the ground.

AEONIUM CANARIENSE hybrid | Crassulaceae | 27°F (-2°C)
Aeonium | Winter-Spring | Canary Islands

The many species of aeoniums are particularly showy near the ocean, thrive in milder inland areas, but will stand little frost. All are striking in form, consisting of rosettes of green or purple leaves frequently borne on branched stems up to several feet high. Aeoniums in the above photo show how well they have naturalized near the ocean in southern California. Their bright yellow cones are especially showy when mixed with aloes (*Aloe striata*), seen in the background of the picture. They are also shown on page 158 upper left behind *Kalanchoe.*

KALANCHOE BLOSSFELDIANA

Crassulaceae	28°F (-2°C)
Winter-Spring	Malagasy

The orange-red flowers in the foreground left are of *Kalanchoe blossfeldiana,* an excellent flowering pot-plant but also useful in the ground for flaming winter and spring color outdoors. In mild areas it provides spectacular color for a long period of time and is particularly striking when contrasted to brilliant yellow-flowered aeoniums as shown here.

CRASSULA FALCATA

Crassulaceae	27°F (-3°C)
Summer-Fall	South Africa

The showy red flowers of this succulent plant have made it a favorite of plant collectors. The attractive flowers keep well when cut and are prized by floral decorators. The unusual gray-bluish sickle-shaped leaves are closely clustered at the base of the stem and provide an interesting contrast with pebbles or colored mulch materials. As with other succulents, watering may be casual.

SEDUM DENDROIDEUM PRAEALTUM | Crassulaceae | 25°F (-4°C)
Spring-Summer | Mexico

Both the species and the variants of this succulent are useful for massed plantings. They have attractive foliage, borne on stems several feet (1 m or more) high. This variety illustrated in the background in the photograph above has green leaves and clear yellow flowers. Little water or care is required. The plants grow readily from cuttings set in the ground and thrive best in sunny locations, in well-drained soil. This species is tender, but other hardier species available make very refined, satisfactory ground covers. They are useful for planting on dry walls.

The photograph also shows *Sedum cupressoides,* in the right foreground, a decumbent, evergreen perennial with cypresslike leaves and bright yellow flowers, blooming in mid- to late summer. Across the path the yucca-like plant with the red flowering stems is *Beschorneria yuccoides* surrounded at the base by gray-leaved *Crassula deltoides. Sedum* X *rubrotinctum* is in the left foreground and is more conspicuous in the retaining wall structure shown on page 160. This is a smaller species noted for the plump rounded leaves which often become ruddy-colored in the sun. Plants may be grown from leaves or stem cuttings. The flowers are yellow. The upright plants growing with the *Sedum* on page 160 are *Kalanchoe tubiflora* from Malagasy.

Sedum X rubrotinctum

AGAVE
Century plant

Agavaceae 29°F (-2°C)
Variable North and South America

Forbidding as their spiky leaves may appear, agaves add a dramatic touch to the subtropical or tropical garden. Their bold rosettes are beautifully symmetrical and seem appropriate to hot and arid areas; illustrated in the right foreground on page 151.

Commonest is *A. americana,* with gigantic clusters of bluish-gray leaves useful in a background planting. Several of its variegated cultivars are even more decorative. Beloved of professional landscapers is *A. attenuata,* the foxtail agave, so named for its arching-pendent brush of flowers. Its leaves are pliable and spineless but easily damaged by frosts. Related and most desirable is *A. vilmoriniana,* with an erect 20 foot (6 m) column of golden yellow flowers. It does not have offset rosettes but dies after flowering producing hundreds of young plants on the inflorescence. *A. filifera,* an excellent rockery plant, has foot-wide (30 cm) clusters of white-streaked, thread-margined leaves and bottlebrush-like flower stalks.

Also small but with wide handsome gray leaves are *A. huachucensis* and *A. shawii* illustrated on this page. Shaw's agave is found in a few areas of California just north of the Mexican border and extends into Baja California. This agave's 9 foot (2.7 m) stem, arising from the typical fleshy-leaved rosette, bears greenish-yellow flowers in open heads sometime between early fall and late spring. This species is most at home in a desert type garden and demands little except well-drained soil.

Despite their common name, most century plants bloom every seven to fifteen years, those species not producing offsets perishing after flowering. Spiny types are best kept away from paths.

ALOE	Liliaceae Winter, Spring, Summer	22°F (-5.5°C) South Africa

The aloes (pictured above, opposite, and on page 157) are succulents that come mostly from various parts of Africa. They exist in a wide range of sizes and colors and provide colorful massed effects useful as ground cover plantings for some sites, as accent plants in others. Most bloom during winter months, but by combining species, the blooming season can be extended. Their inflorescences resemble red-hot pokers with combinations and different shades of red, yellow, and orange.

Aloes require infrequent watering. The old foliage should be cleared out as needed to maintain a neat appearance. Aloes root easily from stem cuttings planted directly in the ground. A notable collection of these plants is in the Huntington Botanical Gardens, San Marino, California. The picture opposite is *A. comptonii,* one of the smaller species. Additional aloes of merit are *A. bainesii,* a tender but impressive tree of 40 feet (12 m); *A. ferox,* slowly reaching 10 feet (3 m) and with candelabras of orange flowers; and the shrubby *A. pluridens* and *A. arborescens,* both with highly colored floral displays. Smaller and for pots and borders are *A. variegata, A.* X *virens* and *A. brevifolia.*

Aloe ciliaris, above, the climbing aloe, is not a true vine but a scrambling shrub, best trained on walls or steep banks to display its orange-red

green-tipped flowers of gemlike brilliance and distinction. A bit slow to start, the plant is probably best with a little shade, especially at its roots. It ultimately achieves 8 feet (2.4 m) or more with periodic burst of flowers in stalked sprays some 8 inches (20 cm) high. Although indifferent to soil, it will not tolerate continued freezing. Prune while flowering or immediately after.

Heliconia rostrata

Heliconia aurantiaca

HELICONIA Lobster claw; false bird-of-paradise	Heliconiaceae Intermittent all year	28°F (-2°C) Tropical America, Pacific Islands

Heliconias are popular garden items in the tropics and warmer areas of the subtropics. They may be used in mass or in borders as well as individual specimens. A number of species are available ranging in height from about 1 foot (30 cm) to 20 feet (6 m) or more. All have decorative foliage and showy bracts in the inflorescence. Inflorescences may be compact or open, erect or pendulous. Few to several yellow or red flowers are borne in each bract and are followed by bright blue berries.

Since the plants can make large clumps, adequate room must be allowed. They thrive in good soil with adequate water. Plants may be propagated by seed or divisions.

HEDYCHIUM GARDNERANUM	Zingiberaceae	28°F (-2°C)
Kahili ginger	Summer	Tropical Asia

Several species of *Hedychium* are commonly cultivated for their showy inflorescences and their fragrance. They thrive in rich soil with adequate moisture and will spread by rhizomes. They are easy to propagate by divisions. *Hedychium gardneranum,* Kahili ginger, has stems to 6 feet (1.8 m) terminating in open inflorescences to 1 foot (0.9 m) long of very fragrant yellow flowers with bright red stamens. It is the hardiest of the species. *Hedychium coronarium,* white ginger, has pure white flowers borne in a compact head. It has naturalized throughout the tropics. *Hedychium flavescens,* yellow ginger, has yellow flowers in a compact head; *Hedychium greenei,* red flowers in a compact head.

RUSSELLIA EQUISETIFORMIS	Scrophulariaceae	30°F (-1°C)
Coral plant; fountain plant	All year	Mexico

This native of Mexico is an attractive shrub to 4 feet (1.2 m) with leafless green drooping branches. The bright red nodding flowers to 1 inch (2.5 m) long are borne along the branches making a spectacular show when used as a bank cover, a low hedge, or in a rockery. It is particularly adaptable to drier areas but has naturalized in many warm wet parts of the world such as the West Indies. The naturally drooping branches make coral plants useful in hanging baskets. They flower best in full sun. Plants are easily propagated by cuttings.

CLIVIA MINIATA
Kaffir lily

Amaryllidaceae
Spring

25°F (-4°C)
South Africa

Masses of Kaffir lily make an attractive green ground cover during summer and winter and in the spring months a spectacular show with clusters of pale orange to scarlet flowers. The evergreen strap-shaped leaves are up to 30 inches (75 cm) long and the flowering stalk is 18-30 inches (45-75 cm) high. Each stalk bears 12 to 20 flowers with each flower up to 3 inches (7.5 cm) long. The fruit is a decorative large red berry. In some areas *Clivia* will naturalize from seed and the clumps will spread from offsets. The plants are best in partial shade and must be protected from slugs and snails.

PACHYSTACHYS LUTEA	Acanthaceae	25°F (-4°C)
	Spring-Fall	Peru

Pachystachys has been used as a container plant but may be grown in the ground for a mass of color. It is shrubby to 3 feet (0.9 m) bearing 4 inch (10 cm) long spikes of white flowers subtended by bright yellow bracts. Flowering occurs throughout the warm months. Propagation is by cuttings.

The related *Pachystachys coccinea,* Cardinal's guard, is noted for the dense terminal spikes of scarlet flowers subtended by green bracts.

V
GROWING CALIFORNIA NATIVE PLANTS

Native plants are often the answer to difficult garden situations and offer a wide choice of plants adapted to local climatic conditions. With increasing development of urban areas often the only way to save a native species from destruction is to bring it into the garden. The interest in native plants has resulted in establishment of botanical gardens devoted to the native flora. In Australia the National Botanical Garden in Canberra displays the wealth of Australian plant material available for cultivation and only a small portion of the plants displayed have been exported. The garden at Kirstenbosch in South Africa, Rancho Santa Ana Botanic Garden, and the Santa Barbara Botanic Garden in southern California are other gardens devoted to the display and introduction of native plants. This chapter will describe a few of the native plants of California, some of which have been introduced into gardens in other parts of the world. In preceding chapters and in the lists in the appendixes the reader will find descriptions and information on native plants from many other countries.

Since the times of the earliest explorers, California has been famous for the beauty of its native plants. But it was Theodore Payne who found some of nature's secrets for successful seed germination and culture. His discovery that matilija poppy seeds need the scorching heat of brush fires to germinate, initiated a technique now used for many recalcitrant seeds.

Through years of observation and experimentation he found most native plants should be set out in the fall because winter rains are crucial. No fertilizer or soil amendments should be used in planting. During the first year, watering every two weeks is usually required. Except for plants whose native habitat is moist or damp, they need sharp drainage especially at the soil surface and little or no watering after plants have become established.

California's annual wildflowers are sown in the fall. Those of perennial wildflowers can be started in the spring in seed flats and will often bloom the following year. Some germinate readily while others need special treatment such as a combination of moisture and low temperature. Such seeds may be mixed with fine moist sand or peat moss and refrigerated for several weeks to break dormancy.

California are especially fortunate that a wide array of flowering plants are well adapted to the state's semiarid climate. Through the years plants from all over the world have been introduced, but meanwhile the native plants have survived in the very environment to which imports must become adapted. Many California natives can be grown away from their natural range but if temperatures or evaporation rates differ greatly from those of their native habitat, their culture should be considered experimental.

ESCHSCHOLZIA CALIFORNICA
California poppy

Papaveraceae — Ground cover
Spring — Annual

When California's spring pageant of wild flowers was first noted by Spanish explorers from the ships off the coast, the California poppy, now the state flower, grew in unbelievable quantities. Its display, much diminished now, still enlivens many areas. The poppy is easily grown from seed. It should be raked in lightly in the fall, then kept moist until the rains begin. Seed sown in midspring will produce summer flowers of somewhat smaller size. It is usually grown as an annual but often reseeds. Though it tolerates a wide variety of soils, it does best in the lighter ones.

The many species of the California poppy vary from large flowers with deep orange hues to small yellow cups with orange tipped petals. Charming garden selections run the gamut of color from white to red and even pink, single and double.

Among the poppies in the landscape above (taken at Santa Barbara Botanic Garden) are baby-blue-eyes or *Nemophilia menziesii*, tidy tips or *Layia platyglossa* ssp. *campestris*, and the brilliant yellow sea dahlia, *Coreopsis maritima*, a close relative of the giant coreopsis shown on page 181.

Arctostaphylos pringlei var. *drupacea*

ARCTOSTAPHYLOS INSULARIS	Ericaceae	Shrub
Island manzanita	Winter	Evergreen

Manzanitas are an important part of hilly California landscapes. Their bright reddish bark on distinctly crooked stems, with green or gray-green ovate leaves that maintain their healthy appearance the year round sets them apart even at a distance. When established, manzanitas are drought resistant and need little or no water during summer months.

Island manzanita, native to Santa Cruz and Santa Rosa islands off the southern California coast, is an erect 6 foot (1.8 m), much-branched shrub, particularly attractive for its smooth shiny elliptical 1 to 2 inch (2.5-5 cm) leaves and red-brown stems. In midwinter or early spring spreading panicles of small waxy white flowers appear followed by yellow-brown fruit called "little apples" by early settlers. This plant is valuable on dry slopes for its year-long color. Propagation by seed is slow and difficult but cuttings can be rooted rather easily.

The pink-bracted manzanita, *Arctostaphylos pringlei* var. *drupacea,* native to southern California mountains is upright growing to 12 feet (3.6 m). Except for the smooth red-brown branches all other parts of the shrub are covered with fine hairs. Delightful fragrant rose-pink flower clusters with pink bracts cover the plant in spring.

Arctostaphylos edmundsii, the Little Sur manzanita, is an attractive low shrub or ground cover for full shade or full sun. Pink flowers are showy in midwinter.

Many other species and named cultivars are available.

BERBERIS (MAHONIA) NEVINII
Nevin's barberry

Berberidaceae
Spring

Shrub
Evergreen

Nevin's barberry, now almost extinct in the wild, has become well established in southern California gardens. It is valuable for its distinctly bluish foliage and yellow flowers in short racemes followed in late summer by reddish berries. It forms a dense shrub 10 feet (3 m) tall with arching branches reaching to the ground. Its growth habit and prickly leaves make it useful as a hedge as well as for shelter and food for birds when considering plants for a sanctuary. This barberry is especially tolerant of drought, being native to hot, sandy areas, but it is also amenable to ordinary garden culture. It may be grown from seed or from nursery stock. The form illustrated, native to the San Fernando Valley, California, was saved from extinction by Theodore Payne.

Berberis (*Mahonia*) *pinnata,* the California mahonia, is a 4 foot (1.2 m) shrub for moist, partly shaded locations. Its fragrant yellow blooms are succeeded by glaucous blue berries.

Berberis (*Mahonia*) *repens,* creeping mahonia, is a good ground cover for partial shade. See page 75 for other mahonias.

Berberis aquifo

CALYCANTHUS OCCIDENTALIS
Spice bush

Calycanthaceae — Shrub
Spring-Summer — Deciduous

Calycanthus occidentalis, known as both spice bush or sweet shrub, is a large spreading shrub to 9 feet (2.7 m). The ovate 2 to 6 inch (5-15 cm) leaves are harsh to the touch but fragrant when crushed. Their rich dark green color makes for a cool lush appearance. Flowers, borne singly at branch ends, are nestled among the foliage. The large, deep red, many petaled blossoms, unique in California native species, appear from spring through midsummer.

Sweet shrub, found throughout the northern coast ranges and Sierra Nevada foothills of California in cool moist canyons, can be used where a background plant is needed in a shaded spot having year-round moisture.

CERCIS OCCIDENTALIS
Western redbud

Leguminosae — Tree-shrub
Spring — Deciduous

Redbud in bloom commands attention in early spring with its mass of red-purple flowers on bare branches making this one of the most showy native trees or large shrubs. Occasionally a white-flowered form is seen. The somewhat divided, heart-shaped leaves are of interest after the flowers are gone. Brown pendent pods appear in fall. Accustomed to dry summers and cold winters, it is a hardy plant and admirably fills the need for a small deciduous tree in thin inland soils. Redbud starts easily from seed but growth is slow. *C. canadensis* (eastern United States) and *C. siliquastrum* (Europe) are 30 foot (9 m) trees with rosy-pink or occasionally white flowers.

CHILOPSIS LINEARIS
Desert willow

Bignoniaceae	Shrub-tree
Spring-Summer	Deciduous

Desert willow is a large spreading shrub or small tree found in desert washes. In late spring and summer large, fragrant, catalpa-like flowers in white to pink or lavender with darker lines grace slender arching branches. The blossoms are followed by thin narrow pods containing many flat seeds with a tuft of hairs at each end. The narrow, willowlike leaves are shed in winter to reveal a framework of limbs with shredded bark. Desert willow should be watered moderately or else grown where loose soil allows the roots to penetrate to a supply of moisture below. It does best inland. Near the coast it should be placed in as warm a location as possible.

FALLUGIA PARADOXA
Apache plume

Rosaceae	Shrub
Spring	Deciduous

Apache plume, native of the pinyon-juniper woodland of the eastern Mojave Desert of California, is a low shrub to 5 feet (1.5 m). In late spring one to one and a half inch (2.5-3.7 cm) white flowers are produced singly at the ends of the branches, followed by small dry fruits with persistent feathery, plumelike styles that give the plant its picturesque name. The many branchlets are covered with a noticeable thin flaky bark. The small leaves are divided into a number of narrow leaflets with incurved edges.

Apache plume, because of its native habitat, is tolerant of temperature extremes and considerable drought and wind.

CHRYSOTHAMNUS	Compositae	Shrub
Rabbit brush	Fall	Deciduous

Chrysothamnus, usually known as rabbit brush, gives to the high desert of California much of its fall color, when these bushy plants with their panicles of showy yellow flowers often dominate the landscape. The small narrow, gray-green leaves are at this time less noticeable compared with the heady show of bloom. The shrub may be from 2 to as much as 10 feet (0.6-3 m) tall. It is quite tolerant of alkali and grows on plains and slopes alike.

Rabbit brush can be used where late season bloom is desired and the fine grayish foliage is acceptable, especially where hot summers and severe winters limit the selection of plant materials.

This can be a valuable plant for erosion control on steep loose banks where summers are warm and dry. These shrubs are frequently used on road fills.

KECKIELLA (PENSTEMON) CORDIFOLIA — Scrophulariaceae — Perennial

Heart-leaved penstemon — Spring-Summer — Semideciduous

Of the nearly 60 penstemons found in California, many of which are handsome garden subjects, this species, now considered a distinct genus, is one of the few that is half-climbing in habit and therefore is sometimes known as "climbing penstemon." The leaves are somewhat heart-shaped, smooth and dark green with strong veining, and up to 2 inches (5 cm) long. The dull scarlet flowers appear in late spring and early summer in compact drooping panicles borne upside down. This plant is often seen draping down roadside banks where a single plant may cover several square yards. Near the coast it prefers a sunny slope but inland it keeps better foliage if shaded for part of the day. It can be propagated from seed or cuttings.

Many species of penstemon are desirable ornamentals. Among the native species recommended are *P. grinnellii,* a spring and summer bloomer with pinkish-white flowers, *P. heterophyllus* with rose-violet flowers in spring and summer, and *P. palmeri* with pinkish-white flowers in spring. These three species are perennial herbs, the first two reaching about one and a half feet (0.5 m) and the latter four feet (1.2 m).

PICKERINGIA MONTANA — Leguminosae — Shrub

Chaparral pea — Spring-Summer — Evergreen

Large reddish-purple pea-shaped flowers in the spring and summer set this shrub apart from the many small-leaved shrubs of the chaparral community. This evergreen plant, 3 to 6 feet (0.9-1.8 m) in height, has stiff spiny branches and olive-green leaves less than a half inch (1.2 cm) in length. Flat 1 to 2 inch (2.5-5 cm) pods are produced only rarely and the plant depends more for propagation on rooting underground stems. It is at home in California on dry slopes of the coast ranges, especially from Santa Barbara County northward.

Under cultivation this shrub is primarily for dry hillside gardens. It has been rather difficult to establish in clay soil.

ZAUSCHNERIA CANA — Onagraceae — Herbaceous perennial

California fuchsia — Summer-Fall — Evergreen

This member of the evening primrose family has long been cultivated for its showy red blossoms that continue from late summer throughout the fall. The flowers are long funnel-form, flaring at the ends and an inch or so (2.5 cm) long. California fuchsia is found from Monterey to Los Angeles counties and on the Channel Islands.

The plant, as well as the more common *Z. californica,* is particularly useful on slopes or in rock gardens as a ground cover where it forms a fairly dense mat 18 inches (45 cm) or more across and perhaps 2 feet (60 cm) high. In the shade it often grows erect. When planted rather closely and pinched back it fills in nicely to produce color at a time when flowers are scarce.

DENDROMECON RIGIDA
Yellow tree poppy

Papaveraceae Shrub
Spring Deciduous

Tree or bush poppy, the only native shrub member of the poppy family, makes a charming addition to drier gardens. Its large, clear yellow flowers are produced over a long season if water is given to supplement rainfall. In fact, some blossoms may be found at any time of the year. This is an open, stiffly branched plant usually between 6 and 10 feet (1.8-3 m) high. The narrow, leathery leaves are up to 4 inches (10 cm) long.

The island counterparts of the mainland species are *Dendromecon rigida harfordii* and *D. rigida rhamnoides,* with slightly larger flowers and broader leaves. The habit of these plants is somewhat less stiff and there is a greater tendency to produce some flowers throughout the year.

Propagation by seed requires sowing in flats then burning over using straw or pine needles, followed by watering. Cuttings may be taken when plants are in full bloom or in summer from ripened wood.

COREOPSIS GIGANTEA	Compositae	Woody perennial
Giant coreopsis	Spring	Deciduous

The giant coreopsis is unique among eleven California native species of this genus. A soft woody trunk up to 4 inches (10 cm) thick and from 3 to 10 feet (0.9-3 m) tall is crowned by several spreading branches. During most of the year the plant is leafless and will remain dormant even with summer watering. Growth resumes at the start of the rainy season when the branches soon become covered with finely dissected light green leaves up to a foot (30 cm) in length. The large 3 inch (7.5 cm) sunflowerlike blooms follow early in the spring, their rich golden yellow adorning the coastal headlands in a striking manner.

FREMONTODENDRON CALIFORNICUM	Sterculiaceae	Shrub
Flannel bush; fremontia	Spring	Evergreen

Fremontia, known also as flannel bush and California slippery elm, is one of the most popular native shrubs. It is a large spreading plant 12 feet (3.6 m) or more in height, and produces in late spring masses of 2 inch (5 cm) flowers of a strong clear yellow. The rounded, lobed leaves are dull green with tawny hairs beneath. It is tolerant of both heat and cold but is very sensitive to excess moisture. Good drainage and sparse summer watering are required.

Fremontodendron mexicanum, a species native to San Diego County and Baja California, has flowers to 3½ inches (8.7 cm) across. This plant is less showy, though it blooms all spring, because the blossoms are partly hidden among the leaves.

Hybrids between the two species have been developed. 'California Glory' is a notable one that has a magnificent spring display with a longer blooming period than the parents with often a second lighter show of blossoms in early summer.

Ceanothus leucodermis

Ceanothus impressus

CEANOTHUS LEUCODERMIS	Rhamnaceae	Shrub
Chaparral whitethorn	Spring	Semideciduous

This large spreading shrub has deep persistent roots that make it valuable in erosion control on dry, stony slopes where other plants would not survive. The stiff and spiny branchlets with small dull green leaves are covered in early spring with clusters of white to pale blue blossoms.

Ceanothus impressus, the Santa Barbara ceanothus, becomes a spreading plant with close-set branches to about 5 feet (15 m) in height. In spring the small deep blue flowers are a pleasing foil against the dark green deeply grooved foliage. Though native on the coast, it withstands the heat of interior valleys. Light sandy soil is preferred but, like all ceanothus, it should be watered sparingly. One of the most popular *Ceanothus* is the evergreen 'Julia Phelps.' It is a compact shrub and eventually forms a mound up to 6 feet (1.8 m) high and 10 (3 m) across with pleasing dark blue flowers.

LEPTODACTYLON CALIFORNICUM
Prickly phlox

Polemoniaceae — Shrub
Spring — Evergreen

Anyone driving through California coast range canyons during spring may see a small shrub hanging from gravelly banks and bearing dense flower heads of deep rose pink. This is the prickly phlox. It not only cliff-hangs but may also be found on hot, well-drained slopes. A 1 to 3 foot (0.3-0.9 m) shrub, it is sometimes straggly, sometimes compact. The clusters of short needlelike leaves are closely set. When out of bloom the shrub is inconspicuous, but the spring display of rich pink is an enjoyable sight.

MIMULUS (DIPLACUS) hybrids
Bush monkey flower

Scrophulariaceae — Shrub
Spring-Summer — Semideciduous

The shrubby species of *Mimulus* are often put in a separate genus *Diplacus* and are among the brightest and longest blooming of all the hillside plants. The red, yellow, orange, buff, and white blossoms persist from early spring through late summer. Tubular flowers terminate in flaring notched or fluted lobes. The shrub is from 1 to 4 feet (0.3-1.2 m) high and has small intricately branched stems with narrow, toothed, glutinous leaves to 3 inches (7.5 cm) long. Hybrids in selected colors are now available. The plants will bloom for months if watered from spring through late summer, then allowed to go dry in the fall.

HETEROMELES (PHOTINIA) ARBUTIFOLIA	Rosaceae	Tree-shrub
Christmas berry; toyon	Summer	Evergreen

The California holly, toyon, or Christmas berry, is perhaps the most widely known of California native shrubs. From late fall to early spring its large clusters of bright red berries are seen on many hillsides. It has been protected by law for many years. The toyon grows to 10 feet (3m) or more high and 15 feet (4.5 m) or more across. With proper pruning it can be developed into a tree. The dark green leathery leaves are narrowly oblong, finely toothed, and about 4 inches (10 cm) long. In early summer large dense clusters of small creamy white blossoms appear. The year-round beauty of this evergreen has earned it a favored spot on lists of natives for general landscape use. In shady spots it becomes tall and leggy and in heavy soils is subject to root rot. Otherwise it is tolerant of summer watering. Native to the Channel Islands is a larger-berried form, the more desirable type for gardens.

CARPENTERIA CALIFORNICA	Saxifragaceae	Shrub
Tree anemone	Spring-Summer	Evergreen

Lovely is the word for *Carpenteria* in bloom. The white anemone-like flowers with inch (2.5 cm) long petals are borne in small clusters at the tips of erect branches. Flowers in late spring and early summer are succeeded by conical capsules with numerous fine seeds. The rather narrow opposite leaves, 3 inches (7.5 cm) or more in length, are dark green above but covered with fine whitish hair on the undersurface.

Tree anemone is found naturally only in Fresno County at lower-middle elevations between the San Joaquin and Kings rivers. It is now frequently seen in cultivation and is at its best in part shade in sandy soil. With some summer watering it retains its good appearance through most of the season.

ROMNEYA COULTERI	Papaveraceae	Perennial
Matilija poppy	Spring-Summer	Deciduous

"Queen of California wild flowers" is the distinction accorded the matilija poppy, whose huge white blossoms make the plant a showpiece from midspring through summer. The size of the flowers is enhanced by the contrast between white crepe paperlike petals and rich yellow stamens, resulting in the often used common name "fried egg plant."

Clustered stems rise from half woody bases to a height of 3 to 7 feet (0.9-2.1 m). The large divided leaves are bluish green and occur the full length of the stems. Plants may be started from seed though germination is somewhat difficult. They may also be propagated during the winter season from root cuttings or by separating parts of an old clump. The roots should not be disturbed more than is necessary but, once established, the matilija poppy is extremely hardy. Plants should be allowed to go dry in late summer and the stems pruned in early winter to within a few inches of the ground.

ENCELIA FARINOSA	Compositae	Shrub
Brittle bush; incienso	Spring	Evergreen

In early to late spring the gravelly mesas and washes of the desert come alive with the brilliant deep yellow blossoms of brittle bush. At this season the erect stems of the 2 to 3 foot (0.6-0.9 m) mound-shaped shrubs are literally covered by daisylike blooms that continue throughout the spring. During the remainder of the year the plant returns to its modest role of melting into the desert landscape, its light gray leaves on brittle stems drawing little attention from the passerby.

Brittle bush is best suited to sandy soil in the interior valleys and deserts. Its seasonal color and continuous fragrance, which prompted the Spanish name "incienso," make it well worth including in the desert garden.

CALIFORNIA GOLD

In the golden fields of wild flowers characteristic of California's springtime, the flower that adds the most widespread touches of gold next to the California poppy is goldfields or sunshine, *Lasthenia chrysostoma.* This is the low growing annual seen in the field. It is a true yellow and so may be distinguished at a distance from the poppy.

Owl's clover, *Orthocarpus purpurascens,* identified by its purplish pink flowers among the goldfields, provides a color not often found in such large masses. The colors of these two plants growing under and around the valley oaks make the picture typical of the California springtime landscape.

LAVATERA ASSURGENTIFLORA	Malvaceae	Shrub
Tree mallow	Spring-Fall	Evergreen

The tree mallow, also known as malva rosa and California windbreak, is native to the Channel Islands of California but has long since been introduced to the coastal mainland. It is a spreading evergreen shrub up to 15 feet (4.5 m) high and at least as broad. The lobed leaves grow to 8 inches (20 cm) across. Flowers are rose-colored with darker veinings and appear continuously from early spring to late fall.

This plant, because of its rapid growth and resistance to buffeting sea winds, has been widely used as a windbreak. It withstands some heat and is considered half hardy.

TRICHOSTEMA LANATUM | Labiatae | Shrub
Woolly blue curls | Spring-Summer | Evergreen

Woolly blue curls, sometime known as romero, is one of the most striking members of the mint family. It is an erect shrub growing from 3 to 5 feet (0.9-1.5 m) tall, with narrow leaves 2 to 3 inches (5-7.5 cm) long, green above and light woolly beneath. The flowers, grouped along the upper part of the stem, are covered with dense woolly blue to purple hairs. Inch-long (2.5 cm) stamens project from the blue flowers in a prominent, almost comical manner. Woolly blue curls is native to the dry hillsides and sandy flats of coastal regions from Monterey south.

This aromatic plant is one of the popular California native shrubs in cultivation where it flowers over a long season. Its requirements are minimal except for the need of good drainage. However, the roots are very sensitive to disturbance during transplanting.

LYONOTHAMNUS FLORIBUNDUS ASPLENIFOLIUS
Catalina ironwood

Rosaceae — Tree
Spring — Evergreen

Lyonothamnus, named for an early Los Angeles forester and horticulturist, is one of the trees endemic to the Channel Islands of California. It is a slender tree under 50 feet (15 m) in height with thin reddish brown bark that shreds off in narrow strips. Compound fernlike leaves give a softened appearance unlike most plants growing in the same environment. Small white flowers, borne in dense clusters at the tips of the branches, are succeeded by small capsules. This variety is most commonly grown because of its finely notched foliage. It should be given good drainage and watered moderately.

AESCULUS CALIFORNICA
Buckeye; horse-chestnut

Hippocastanaceae — Tree
Summer — Deciduous

In the month of June, travelers in the foothills cannot help noticing a small tree or large shrub with prominent spikes of white flowers. This is the California buckeye and the state's only member of the buckeye family. It is found from Shasta and Siskiyou counties to northern Los Angeles County below 4,000 feet (1,200 m) in the Sierra Nevada and coast ranges.

Typically it is a round-topped tree of 15 to 20 feet (4.5-6 m). The showy compound leaves consist of leaflets 2 to 6 inches (5-15 cm) in length. The irregularly shaped flowers are about half an inch (1.2 cm) in length, hundreds of them clothing the ample erect or rarely pendulous spikes.

The trees are early deciduous, starting in midsummer to reveal the light gray trunk and branches. This plant can be used where summer drought resistant specimens are needed and the early fall of leaves is acceptable.

PSOROTHAMNUS SPINOSA | Leguminosae | Tree
(*Dalea spinosa*)
Smoke tree | Summer | Deciduous

The smoke tree of the low desert has always had a special appeal to artists and others sensitive to the beauty of that area. When seen from a distance the tree's smoky gray branches blend into the landscape, making it appear to be a link between the plant world and the atmosphere. In early summer it puts forth its inch (2.5 cm) long spikes of rich blue-violet flowers that cover the entire plant.

As a tree, this species may reach a height of 18 feet (5.4 m) but it frequently grows as a large spreading shrub. A coarse soil and good drainage appear to be necessary since this plant is almost exclusively confined to sandy washes. It is recommended as a garden subject only for desert areas. The hard seed coats are said to be scoured by the tumbling sand and gravel during storms, after which the seed germinates more readily.

Calochortus venustus

Calochortus venustus

Calochortus clavatus

Calochortus kennedyi

CALOCHORTUS	Liliaceae	Bulb
Mariposa lilies	Spring	Deciduous

Mariposa lilies excite the admiration of all who see them. Many colors are represented in the group but none are quite so intense as the orange to vermillion shades of the desert mariposa (*C. kennedyi*). The plant bears one to several broad cup-shaped flowers on stems from 5 to 20 inches (5-50 cm) high with a few narrow grasslike leaves arising from the ground. A native of the high deserts, this species is highly questionable as a garden subject.

Other species such as *C. catalinae, C. luteus, C. venustus, C. weedii, C. clavatus,* and *C. vestae,* ranging in color from yellow and pink to white, are much more suitable but should be given excellent drainage and protection from rodents, especially gophers. The bulbs must remain dry during the summer.

The most suitable mariposa lily for garden culture is *C. vestae.* It may be grown from seed sown in flats in the fall and left until the following season before transplanting. Mariposa lilies are more easily grown from bulbs placed in a special bed with good drainage.

FOUQUIERIA SPLENDENS	Fouquieriaceae	Shrub
Ocotillo	Spring	Deciduous

Ocotillo is one of the desert's oddities. Most of the year its cluster of spiny canes appears almost lifeless. Soon after sufficient rain, small leaves appear and in spring the stems become tipped with panicles of scarlet flowers, giving the plant the appearance of lighted tapers. Within a few weeks the leaves and blossoms fall and the plant returns to dormancy.

Although most at home in the warmer deserts, ocotillo can be grown almost anywhere in the southern California lowlands as long as good drainage is provided and too much moisture is avoided. Plants may be started by simply cutting off a portion of a cane, which is dried for a few weeks and then set a few inches into the ground. Like most other desert plants, ocotillo is protected by law and should not be removed from public land.

YUCCA

In the late spring the chaparral country assumes a character unknown to it during the rest of the year. It is then that the yuccas send forth fast-growing flower stalks that culminate in a myriad of waxy cream-white blooms. The candles of bloom held aloft to the sky attract a host of insects of which the most important is the *Pronuba* moth that fertilizes the flowers and ensures the plant's survival. *Yucca whipplei,* shown opposite and described on the following page, has several subspecies, some of which form a number of leaf rosettes. The flowering rosette dies at the end of its blooming period. Other types, which form only one group of leaves, die completely after blooming but the thousands of flat black seeds give promise of another generation.

Yucca brevifolia, Joshua tree, ancient monarch of California deserts shown above, is a famous native. Little known is the fact that it belongs to the genus *Yucca.* Joshua trees form a unique woodland covering many square miles in the Mojave Desert. In spring their grotesque forms produce tight clusters of white flowers at the tips of the branches, while underneath is usually found a colorful carpet of desert annuals. They are at home only in the desert garden where they can dominate the scene and provide a setting for many smaller desert plants. They need the summer heat of the interior for vigorous growth.

NOLINA WOLFII
Beargrass

Agavaceae
Spring

Shrublike plant
Evergreen

Nolina resembles yucca with its showy flower stalk and myriads of blossoms. The individual blooms, however, are smaller than those of yucca and the male and female flowers are on separate plants. The flowers tend to be a deeper cream color. *Nolina wolfii* (right), native of inland desert mountains, with a flower stalk that may reach 12 feet (3.6 m), provides a magnificent accent in its natural setting and a striking addition to the desert garden. Additional water will increase its rate of growth.

Parry's nolina, *N. parryi* (left), is found both on desert slopes and in drier parts of the chaparral areas of southern California. The leaves are narrower and the flowering stalk about half the height of that of *N. wolfii.*

CERCIDIUM FLORIDUM
Palo verde

Leguminosae — Tree
Spring-Summer — Deciduous

Palo verde, Spanish for "green wood," is a tree of the desert washes in the Colorado desert. Most of the year the tree is bare. Its small compound leaves are produced only after rains. During dormant periods the function of the leaves is partly taken over by the blue-green stems and trunk. Sometime during the spring, as if to compensate, there is a notable display of small yellow blossoms on the spiny branchlets. Here again is an example of the predominance of yellow among California's native flowering plants. This tree withstands temperatures considerably below freezing for short periods.

YUCCA WHIPPLEI
Our Lord's Candle

Agavaceae — Shrublike plant
Spring-Summer — Evergreen

Yucca whipplei (page 197) in bloom is an accent plant of stunning beauty. In late spring or early summer a flower stalk bearing hundreds of pendent creamy white flowers grows with astonishing rapidity to a height of 5 to 12 feet (1.5-3.6 m) out of the basal rosette of gray-green finely saw-toothed leaves. Later the blossoms mature into dry spongy capsules containing many flat black seeds.

Yucca is extremely persistent. After fires it sprouts new leaves, often from a charred and seemingly lifeless stump. This is a plant for the dry garden or an "impossible" slope but should be placed out of range of contact with its sharply pointed leaves. It may be grown from seed, with patience, or purchased from a nursery dealing with California native plants.

Lupinus nanus with *Orthocarpus purpurascens*

LUPINUS ARBOREUS
Bush lupine

Leguminosae — Shrub
Spring-Summer — Evergreen

The wildflower fields of California are often filled with lupines of varying sizes and flower color. *Lupinus nanus,* an annual, is cultivated in England and the Russell lupines, derived from *L. polyphyllus,* are well known in horticulture. The shrubby lupines also deserve cultivation although they are often short-lived—three to five years—and must be replanted. Several flower colors are available—clear yellow, light blue, and shades of rose, lavender, and purple. Well-drained soil, sun, and minimal watering are required.

Lupinus arboreus

NEMOPHILA MENZIESII	Hydrophyllaceae	Annual
Baby blue eyes	Spring	

An old favorite in English gardens where it was introduced over 100 years ago this annual, as well as the related Five-spot, *Nemophila maculata,* reseeds and spreads. The blue and white flowers make it an attractive ground cover for moist areas in full sun or partial shade.

LIMNANTHES DOUGLASII | Limnanthaceae | Annual
Meadow foam | Spring

Rarely cultivated in California meadow foam was another early introduction into England where it has escaped from gardens. Particularly useful along garden streams, the mass of white to yellow 1 inch (2.5 cm) flowers gives it the appropriate common name.

CAMISSONIA CHEIRANTHIFOLIA	Onagraceae	Annual or perennial
Beach primrose	Spring-Summer	

The beach primrose with its bright yellow flowers fading orange is one of a large group of species suitable for garden culture. Flowering in full sun in the early spring, it will continue to grow and flower during the summer if moisture is provided. In favorable sites this annual will persist more than one season.

OENOTHERA CAESPITOSA — Onagraceae — Perennial
Evening primrose — Spring-Summer

Flowers open in the late afternoon and remain open during the night attracting their pollinator hawk moths. Mats of silver gray foliage and white flowers, fading pink, provide a fine garden display in full sun and sandy soil. As with many native desert plants minimal watering is needed.

CLARKIA WILLIAMSONII — Onagraceae — Annual
Farewell to Spring — Early summer

The genus *Clarkia* has been known in horticulture for over one hundred years with several species in cultivation. The garden godetias have been favorite items for the florist trade. Horticultural selections are available of *Clarkia unguiculata, C.* (*Godetia*) *rubicunda* and *C.* (*Godetia*) *amoena whitneyi. Clarkia williamsonii,* with flowers varying in color from white with lavender to wine-red, is one of the showy species useful for the garden.

LATHYRUS SPLENDENS | Leguminosae | Perennial vine
Pride of California; Campo pea | Spring-Summer | Evergreen

The perennial Campo pea with its brilliant wine-red to maroon flowers climbs rampantly by tendrils to as much as 8 feet (2.4 m) over fences, walls, and trellises. It is without question the showiest of California native vines and a fine addition to the garden where it thrives in coarse, well-drained soil in sunny dry locations. Plants may be grown from seed or cuttings but are difficult to transplant. Once established they should never be watered in summer.

Dudleya cymosa ssp. *ovatifolia*

DUDLEYA species
Live-for-ever

Crassulaceae	Succulent
Spring-Summer	Evergreen

Dudleya, with a number of California native species, is a useful cultivated plant—a perennial ground cover massed in partial shade, in a rockery, a single rosette carefully sited on a wall, or showy in a container. The foliage varies from green with occasional reddish or brownish tints to chalky white. Flower color may be greenish, white, yellow, or red. Fast drainage is essential and overhead watering should be avoided.

Dudleya pulverulenta

CACTI

Cacti account for some of the most distinctive native plants and are found throughout California. Many times they occur in company with yuccas, ocotillos, and other xerophytes. Under cultivation cacti grow and thrive in most areas except where extremely cold or humid. The desert types will withstand some frost. Good drainage is essential and it is safest to plant them in beds raised well above the immediate surroundings. The bloom occurs irregularly from early to late spring.

The best source for cacti is a nursery dealing in these plants but those willing to devote the necessary time can grow them from seed.

ECHINOCEREUS ENGELMANNII	Cactaceae	Succulent
Hedgehog cactus	Spring	Evergreen

The hedgehog cactus forms low clumps of erect cylinder-shaped branches arising from the ground. Each branch has a dozen or so ribs and is set with stout spines in a variety of colors: red, white, gray, yellow, and brown. Its appearance has prompted the name cucumber cactus. The flowers are crimson-magenta. This plant is found in coarse soils on slopes and benches in many locations in the California deserts.

OPUNTIA LITTORALIS
Prickly pear; tuna cactus

Cactaceae — Succulent
Spring — Evergreen

Prickly pear is the name given to the many species of the genus *Opuntia* with flattened joints and juicy fruits which were relished by the Indians and are still sold in some markets of the southwest. *Opuntia littoralis* forms a many-branched shrub up to 3 feet (0.9 m) or more in height. While most frequently used as an accent shrub in gardens, in many older gardens it was sometimes used as a fence. In time the prickly pear can provide a fire retardant barrier of considerable value. In late spring attractive yellow blossoms appear around the edge of its flattened joints. It is easily propagated from cuttings placed directly in sandy, well-drained soil, and it requires little water.

OPUNTIA BASILARIS
Beavertail cactus

Cactaceae — Succulent
Spring — Evergreen

One of the common cacti of the California deserts deserves a place in any dry garden. The showy cerise to rose, yellow or rarely white flowers are 2-3 inches (5-7.5 cm) across, clustered at the upper ends of the green, gray, or often purplish joints. The plants are low and spreading. Beavertail cactus occurs natively not only in California but extends northeast to Utah and south into Mexico.

FEROCACTUS ACANTHODES — Cactaceae — Succulent
Barrel cactus — Spring — Evergreen

The well-known barrel cactus was once a familiar sight on desert slopes. Now, though protected by law, it has been largely removed from the more accessible sites. When young it is globular in shape but in time assumes the typical barrel form. A ring of yellow flowers appears in spring at the crown of the plant. The size and form of this cactus with its spiny, fluted sides, are its salient features. Although it contains a great deal of moisture, the sap is unpalatable, a very poor substitute for a drink of clear water.

OPUNTIA FICUS-INDICA	Cactaceae	Succulent
Indian fig	Spring	Evergreen

The origin of the Indian fig is unknown but it is so common in old California gardens that many think it a native. It is widely grown throughout the tropics and subtropics and its introduction into California dates from the early Mission days. Indian fig may reach 18 feet (5.4 m) in height. The 3-4 inch (7.5-10 cm) yellow flowers are followed by edible fruits that may vary in color from white to red. Burbank's Spineless Cactus is a selection from this species.

APPENDIXES

APPENDIX I
ADDITIONAL FLOWERING TREES OF MERIT

EVERGREEN TREES (or shortly deciduous, in colder areas may be completely deciduous. Flowering for many of the more tropical species is during the dry season.)

FLOWERS WHITE

Angophora costata gum myrtle	Myrtaceae Australia	40-50 ft. (12-15 m)	summer	25°F (-4°C)
Arbutus unedo strawberry tree	Ericaceae S. Europe	20-30 ft. (6-9 m)	autumn-winter fruits red	25°F (-4°C)
Barringtonia asiatica	Barringtoniaceae Madagascar-Pacific Islands	15 ft (4.5 m)	spring-summer stamens pink	30°F (-1°C)
Clethra arborea lily of the valley tree	Clethraceae Madeira	20 ft. (6 m)	late summer	24°F (-4.5°C)
Crinodendron patagua lily of the valley tree	Elaeocarpaceae Chile	25 ft. (7.5 m)	summer-autumn	22°F (-5.5°C)
Cunonia capensis African red alder	Cunoniaceae S. Africa	25 ft. (7.5 m)	summer	20°F (-7°C)
Dillenia indica elephant apple	Dilleniaceae Borneo-Java	60 ft. (18 m)	summer	28°F (-2°C)
Dombeya tiliacea Natal cherry	Byttneriaceae S. Africa	25 ft. (7.5 m)	autumn	26°F (-3°C)
Drimys winteri Winter's bark	Winteraceae S. America	25 ft. (7.5 m)	spring	20°F (-7°C)
Elaeocarpus reticulatus blueberry ash	Elaeocarpaceae E. Australia	60 ft. (18 m)	spring-summer fruit blue	25°F (-4°C)
Gordonia axillaris false camellia	Theaceae S. China-Taiwan	10 ft. (3 m)	winter	24°F (-4.5°C)
Hoheria populnea lacebark	Malvaceae New Zealand	30 ft. (9 m)	autumn	22°F (-5.5°C)
Inga spp.	Leguminosae Trop. America	30 ft. (9 m)	intermittent	30°F (-1°C)
Leucaena leucocephala white popinac	Leguminosae Trop. America	30 ft. (9 m)	summer	28°F (-2°C)
Luehea spp.	Tiliaceae Trop. America	25-90 ft. (7.5-27 m)	summer	28°F (-2°C)
Mesua ferrea ironwood tree	Guttiferae Thailand, Malaysia	60 ft. (18 m)	spring	32°F (0°C)
Millingtonia hortensis Indian cork tree	Bignoniaceae Burma, Malaysia	80 ft. (2.4 m)	spring-autumn	20°F (-7°C)
Oncoba spinosa	Flacourtiaceae Trop. Africa	20 ft. (6 m)	summer	30°F -1°C
Pithecellobium dulce Manila tamarind	Leguminosae Mexico-C. America	60 ft. (18 m)	winter	28°F (-2°C)
Pittosporum rhombifolium Queensland pittosporum	Pittosporaceae Australia	35 ft. (10.5 m)	fall, winter showy orange fruit	25°F (-4°C)
Pittosporum undulatum Victorian box	Pittosporaceae Australia	40 ft. (12 m)	spring	25°F (-4°C)
Posoqueria spp. needle flower tree	Rubiaceae Trop. America	20 ft. (6 m)	summer-winter	28°F (-2°C)
Pterospermum acerifolium	Byttneriaceae India-Java	100 ft. (30 m)	midwinter	20°F (-7°C)
Pyrus kawakamii evergreen pear	Rosaceae Taiwan	30 ft. (9 m)	winter-spring	20°F (-7°C)
Schima wallichii	Theaceae India-Taiwan	120 ft. (36 m)		10°F (-12°C)

Syzygium jambos	Myrtaceae	30 ft.	autumn-winter	30°F
rose apple	East Indies	(9 m)		(-1°C)
Syzygium paniculatum	Myrtaceae	40 ft.	autumn-winter	26°F
Australian cherry	Australia	(12 m)	purplish-red fruit	(-3°C)
Talauma hodgsonii	Magnoliaceae	30 ft.	late spring	25°F
	Himalayas	(9 m)		(-4°C)
Tristania conferta	Myrtaceae	60 ft.	summer	26°F
Brisbane box	Australia	(18 m)		(-3°C)

FLOWERS YELLOW TO ORANGE

Banksia grandis	Proteaceae	40 ft.	spring-summer	28°F
giant banksia	W. Australia	(12 m)	yellowish	(-2°C)
Barklya syringifolia	Leguminosae	60 ft.	summer	28°F
gold-blossom tree	NE Australia	(18 m)	golden-yellow	(-2°C)
Caesalpinia spinosa	Leguminosae	12 ft.	winter-spring	20°F
	Cuba, S. America	(3.6 m)	yellow-red	(-7°C)
Cassia brewsteri	Leguminosae	40 ft.	spring	28°F
cigar cassia	Queensland	(12 m)	golden to deep red	(-2°C)
Geissois benthamii	Cunoniaceae	100 ft.	summer	28°F
red carrabeen	Queensland	(30 m)	yellow	(-2°C)
Michelia champaca	Magnoliaceae	80 ft.	intermittent	28°F
golden champa	Himalayas	(24 m)	yellow-orange	(-2°C)
Peltophorum pterocarpum	Leguminosae	50 ft.	spring-summer	26°F
yellow poinciana	SE Asia-Australia	(15 m)	yellow	(-3°C)
Pithecellobium flexicaule	Leguminosae	50 ft.	spring-summer	28°F
Texas ebony	S. Texas-NE Mexico	(15 m)	yellow	(-2°C)
Pittosporum phillyraeoides	Pittosporaceae	20 ft.	spring	25°F
weeping pittosporum	Australia	(6 m)	yellow	(-4°C)
Schizolobium parahybum	Leguminosae	100 ft.	spring-summer	30°F
	Trop. America	(30 m)	light yellow	(-1°C)
Thespesia populnea	Malvaceae	60 ft.	intermittent	30°F
tulip tree, portia tree	Pantropical	(18 m)	yellow-orange-yellow	(-1°C)
Viminaria denudata	Leguminosae	20 ft.	early summer	28°F
	SE Australia	(6 m)	orange-yellow	(-2°C)
Warszewiczia coccinea	Rubiaceae	40 ft.	intermittent	40°F
wild poinsettia, bandera	Trop. America	(12 m)	yellow-orange	(7.5°C)
			enlarged red calyx lobe	
Wercklea lutea	Malvaceae	25 ft.	summer	35°F
	Costa Rica	(7.5 m)	yellow	(2°C)

FLOWERS PINK TO RED

Amherstia nobilis	Leguminosae	40 ft.	intermittent	65°F
pride of Burma	Burma	(12 m)	bright red and yellow	(18°C)
Andira inermis	Leguminosae	60 ft.	spring	24°F
almendro	Trop. America-W. Africa	(18 m)	lilac-pale rose	(-4.5°C)
Barringtonia acutangula	Barringtoniaceae	40 ft.	spring-summer	30°F
Indian oak	India-N. Australia	(12 m)	red	(-1°C)
Barringtonia racemosa	Barringtoniaceae	50 ft.	spring-summer	30°F
	Malaysia-Pacific Islands	(15 m)	white-red	(-1°C)
Brassaia actinophylla	Araliaceae	20 ft.	summer	25°F
Queensland umbrella tree	Australia	(6 m)	red	(-4°C)
Brownea grandiceps	Leguminosae	60 ft.	intermittent	30°F
rose of Venezuela	Venezuela	(18 m)	bright red	(-1°C)
Brownea macrophylla	Leguminosae	30 ft.	spring-summer	35°F
	Panama, Colombia	(9 m)	fire-red	(2°C)
Calotropis gigantea	Asclepiadaceae	15 ft.	summer	28°F
giant milkweed	India-Indonesia	(4.5 m)	rose-purple	(-2°C)
Ceratopetalum gummiferum	Cunoniaceae	40 ft.	summer	28°F
Australian Christmas bush	New South Wales	(12 m)	white-rose	(-2°C)
Clusia rosea	Guttiferae	50 ft.	summer	30°F
Scotch attorney, balsam apple	Trop. America	(15 m)	pink and white	(-1°C)

Cordia sebestena geiger tree	Boraginaceae Trop. America	30 ft. (9 m)	intermittent orange-scarlet	30°F (-1°C)
Couroupita guianensis cannonball tree	Lecythidaceae Trop. America	50 ft. (15 m)	intermittent red-rose-yellow	30°F (-1°C)
Embothrium coccineum Chilean fire tree	Proteaceae Chile-Argentina	50 ft. (15 m)	spring-summer scarlet	12°F (-10°C)
Hakea laurina sea urchin tree	Proteaceae Australia	30 ft. (9 m)	fall and winter cream and red	20°F (-7°C)
Kleinhovia hospita	Byttneriaceae Trop. Africa	60 ft. (18 m)	summer red, yellow, and pink	35°F (1.5°C)
Lagunaria patersonii primrose tree	Malvaceae Australia	40 ft. (12 m)	summer pale rose	25°F (-4°C)
Oreocallis wickhamii red silky oak	Proteaceae NE Australia	30 ft. (9 m))	spring-summer orange-crimson	28°F (-2°C)
Pongamia pinnata karum tree	Leguminosae Trop. Asia, Australia	40 ft. (12 m)	spring white, pink, purplish	28°F (-2°C)
Rhodoleia championii silk rose	Hamamelidaceae China, Hong Kong	20 ft. (6 m)	winter-spring rosy-ruby red	28°F (-2°C)
Rhodosphaera rhodanthema yellowwood	Anacardiaceae Australia	70 ft. (21 m)	spring red	28°F (-2°C)
Samanea saman rain tree, monkey-pod	Leguminosae Trop. America	80 ft. (24 m)	spring-summer yellowish-pink	40°F (4.5°C)
Saraca indica asoka	Leguminosae India-Malaysia	30 ft. (9 m)	intermittent orange-red	30°F (-1°C)
Schotia brachypetala tree fuchsia	Leguminosae S. Africa	20 ft. (6 m)	spring crimson	30°F (-1°C)
Schotia latifolia forest boerboom	Leguminosae S. Africa	30 ft. (9 m)	spring pink or rose	30°F (-1°C)
Schrebera alata	Oleaceae S. Africa	20 ft. (6 m)	late spring pink	25°F (-4°C)
Securidaca longipedunculata Rhodesian violet tree	Polygalaceae Trop. Africa	25 ft. (7.5 m)	early summer red-purple to pink	33°F (1°C)
Sesbania grandiflora scarlet wistaria tree	Leguminosae Trop. Asia	40 ft. (12 m)	spring-summer-winter red-white-pink	30°F (-1°C)
Stereospermum kunthianum pink jacaranda	Bignoniaceae Africa	30 ft. (9 m)	early spring pale pink-lilac	24°F (-4.5°C)
Syzygium malaccense Malay apple	Myrtaceae Malaysia	40 ft. (12 m)	intermittent purplish-red	32°F (0°C)
Virgilia capensis keurboom	Leguminosae S. Africa	30 ft. (9 m)	spring to early winter mauve-pink	26°F (-3°C)
Virgilia divaricata choice tree	Leguminosae S. Africa	30 ft. (9 m)	spring rosy pink	25°F (-4°C)
Vitex lucens chaste tree	Verbenaceae New Zealand	30 ft. (9 m)	winter dark red	25°F (-4°C)
Wercklea insignis	Malvaceae Costa Rica	60 ft. (18 m)	summer-winter lilac rose to pink	35°F (2°C)

FLOWERS LAVENDER, PURPLE, BLUE

Belotia grewiifolia moho, capulín	Tiliaceae Trop. America	60 ft. (18 m)	spring violet with pink sepals	35°F (2°C)
Guaiacum officinale lignum-vitae	Zygophyllaceae Trop. America	30 ft. (9 m)	spring-summer blue	28°F (-2°C)
Guaiacum sanctum lignum-vitae	Zygophyllaceae Trop. America	30 ft. (9 m)	spring-summer blue-purple	30°F (-1°C)
Lysidice rhodostegia	Leguminosae SE China-Vietnam	25 ft. (7.5 m)	spring-autumn bracts rose, flowers violet	28°F (-2°C)
Napoleona imperialis Napoleon's hat	Barringtoniaceae Nigeria	10 ft. (3 m)	spring blue and rose-yellow	28°F (-2°C)
Robinsonella cordata blue hibiscus tree	Malvaceae Guatemala	20 ft. (6 m)	spring violet blue	25°F (-4°C)
Solanum macranthum potato tree	Solanaceae Brazil	30 ft. (9 m)	intermittent bluish violet to bluish-white	26°F (-3°C)
Sophora secundiflora mescal bean	Leguminosae Mexico, N. Mexico, Texas	25 ft. (7.5 m)	spring violet blue	17°F (-8°C)

DECIDUOUS TREES

FLOWERS WHITE

Cornus nuttallii western dogwood	Cornaceae W. United States	75 ft. (22.5 m)	spring	15°F (-9°C)
Davidia involucrata handkerchief tree, dove tree	Nyssaceae China	50 ft. (15 m)	spring white bracts	0°F (-18°C)
Halesia carolina silverbell, snowdrop tree	Styracaceae SE United States	30 ft. (9 m)	spring	10°F (-12°C)
Ipomoea arborescens palo blanco, morning glory tree	Convolvulaceae Mexico-Guatemala	20 ft. (6 m)	winter	30°F (-1°C)
Moringa pterygosperma horseradish tree	Moringaceae India	30 ft. (9 m)	intermittent	30°F (-1°C)
Plumeria rubra acutifolia pagoda tree	Apocynaceae Mexico, W. Indies	15 ft. (4.5 m)	summer-autumn	30°F (-1°C)

FLOWERS YELLOW TO ORANGE

Cassia fistula golden shower	Leguminosae India	30 ft. (9 m)	spring-summer pale yellow	25°F (-4°C)
Cochlospermum vitifolium	Cochlospermaceae Mexico-N. and S. America	40 ft. (12 m)	spring bright yellow	30°F (-1°C)
Crataeva religiosa spider tree, sacred garlic pear	Capparaceae Old World tropics	20 ft. (6 m)	spring-autumn white to yellow-purplish	30°F (-1°C)
Liriodendron tulipifera tulip tree	Magnoliaceae E. United States	75 ft. (22.5 m)	summer green, yellow, and orange	10°F (-12°C)
Parkinsonia aculeata Jerusalem thorn	Leguminosae Trop. America	30 ft. (9 m)	intermittent yellow	15°F (-9°C)
Pterocarpus indicus bloodwood, Burmese rosewood	Leguminosae SE Asia, Philippines	80 ft. (24 m)	summer yellow	35°F (1.5°C)

FLOWERS PINK TO RED

Acrocarpus fraxinifolius	Leguminosae India	60 ft. (18 m)	early spring scarlet	30°F (-1°C)
Aesculus X *carnea* red horse-chestnut	Hippocastanaceae Garden hybrid	40 ft. (12 m)	spring flesh to rosy-pink	10°F (-12°C)
Aesculus X *carnea* 'Briotii' red horse-chestnut	Hippocastanaceae Garden hybrid	40 ft. (12 m)	spring rosy crimson	10°F (-12°C)
Bombax ceiba silk-cotton tree	Bombacaceae India-Malaya	75 ft. (22.5 m)	spring-summer red	28°F (-2°C)
Butea monosperma flame of the forest	Leguminosae India to Burma	50 ft. (15 m)	spring bright orange-red	28°F (-2°C)
Cassia grandis pink shower	Leguminosae Trop. America	50 ft. (15 m)	late winter-spring yellow to salmon and rose-pink	28°F (-2°C)
Cassia javanica apple-blossom cassia	Leguminosae Indonesia	20 ft. (6 m)	spring-summer pale-dark red to rose-pink	25°F (-4°C)
Cassia moschata bronze shower	Leguminosae Trop. America	30 ft. (12 m)	spring yellow-brick-red	30°F (-1°C)
Cercis canadensis redbud	Leguminosae E. United States	30 ft. (12 m)	spring rosy pink to white	(0°F) (-18°C)
Cercis siliquastrum Judas tree	Leguminosae S. Europe	25 ft. (7.5 m)	spring rosy pink	15°F (-9°C)
Cornus florida flowering dogwood	Cornaceae E. United States	40 ft. (12 m)	spring pink, rose, white	0°F (-18°C)

Covillea racemosa	Leguminosae Madagascar	50 ft. (15 m)	summer-autumn orange-scarlet	28°F (-2°C)
Dais cotinifolia pompon tree	Thymelaeaceae S. Africa	20 ft. (6 m)	late spring lavender-pink	25°F (-4°C)
Gliricidia sepium madre de cacao	Leguminosae Trop. America	30 ft. (9 m)	intermittent rose-pink to white	26°F (-3°C)
Kigelia pinnata sausage tree	Bignoniaceae Trop. Africa	50 ft. (15 m)	summer-autumn claret	26°F (-3°C)
Lagerstroemia speciosa pride of India queen's crape myrtle	Lythraceae SE Asia to Australia	80 ft. (24 m)	spring-summer rose-pink, purple to white	25°F (-4°C)
Melia azedarach pride of India	Meliaceae Asia	50 ft. (15 m)	late spring, summer lavender-pink	10°F (-12°C)
Melia azedarach 'Umbraculifera' Texas umbrella tree	Meliaceae	30 ft. (9 m)	late spring, summer pink	10°F (-12°C)
Pachira aquatica water chestnut	Bombacaceae Mexico-S. America	60 ft. (18 m)	intermittent yellowish-white to pink	28°F (-2°C)
Pachira insignis wild chestnut	Bombacaceae Brazil	90 ft. (27 m)	summer-winter brownish-red to scarlet	25°F (-4°C)
Pseudobombax ellipticum shaving-brush tree	Bombacaceae Mexico-C. America	30 ft. (9 m)	spring-summer white-pink	28°F (-2°C)
Pseudobombax grandiflorum	Bombacaceae Brazil	25 ft. (7.5 m)	summer deep purple-red	30°F (-1°C)
Sesbania punicea glory pea	Leguminosae Argentina	12 ft. (3.6 m)	spring-summer scarlet	25°F (-4°C)
Triplaris americana long John	Polygonaceae Trop. America	30 ft. (9 m)	spring white to red in fruit	28°F (-2°C)
Triplaris surinamensis long John	Polygonaceae Trop. America	60 ft. (18 m)	autumn white to red in fruit	28°F (-2°C)

FLOWERS BLUE, VIOLET TO PURPLE

Bolusanthus speciosus Rhodesian wistaria	Leguminosae S. Africa	20 ft. (6 m)	spring, summer violet	20°F (-7°C)
Lonchocarpus nitidus	Leguminosae Brazil	25 ft. (7.5 m)	summer violet	26°F (-3°C)
Lonchocarpus violaceus	Leguminosae West Indies	20 ft. (6 m)	summer white to pale purple and pinkish	25°F (-4°C)

APPENDIX II
ADDITIONAL FLOWERING SHRUBS OF MERIT

This list includes some vining or sprawling shrubs that can be trained on a wall or allowed to droop over a bank, designated here as VS. Small shrubs which may have partially herbaceous stems are designated SS. Shrubs are arranged by flower color. In some cases the same species may have cultivars or variants with a gamut of color. Consequently the arrangement here is somewhat arbitrary and the species are listed under the most common color with an indication of the variations that occur.

FLOWERS WHITE

Abelia X *grandiflora* (pink and white) glossy abelia	Caprifoliaceae Garden	8 ft. (2.4 m)	summer sun, part shade	hardy
Abutilon hybridum (pink, red, yellow, apricot) flowering maple	Malvaceae Garden	10 ft. (3 m)	all year part shade	25°F (-4°C)
Acokanthera oblongifolia (pink and white) bushman's poison	Apocynaceae Africa	10 ft. (3 m)	all year sun, part shade	25°F (-4°C)
Berzelia lanuginosa (cream)	Bruniaceae S. Africa	6 ft. (1.8 m)	spring, summer	28°F (-2°C)
Bouvardia longiflora 'Albatross' SS	Rubiaceae Garden	3 ft. (0.9 m)	spring-autumn sun, part shade	light frost
Brugmansia X *candida* angel's trumpet	Solanaceae Peru	15 ft. (4.5 m)	summer, autumn sun	light frost
Brugmansia suaveolens angel's trumpet	Solanaceae Brazil	15 ft. (4.5 m)	summer sun	light frost
Buddleia asiatica white butterfly bush	Loganiaceae China, India	6 ft. (8 m)	winter-spring sun, good drainage	tender
Cestrum diurnum day jessamine	Solanaceae West Indies	15 ft. (4.5 m)	summer sun, part shade	25°F (-4°C)
Choisya ternata Mexican orange	Rutaceae Mexico	8 ft. (2.4 m)	spring sun, part shade	15°F (-9°C)
Clerodendrum philippinum (white or pink) glory bower	Verbenaceae China, Japan	8 ft. (2.4 m)	summer sun, part shade	light frost
Coffea arabica coffee	Rubiaceae Trop. Africa	15 ft. (4.5 m)	spring-summer part shade	28°F (-2°C)
Coleonema album breath of heaven	Rutaceae S. Africa	5 ft. (1.5 m)	autumn to spring sun, drainage	24°F (-4.5°C)
Cordia boissieri	Boraginaceae Texas, Mexico	10 ft. (3 m)	spring-summer sun	26°F (-3°C)
Correa alba Australian fuchsia	Rutaceae Australia	4 ft. (1.2 m)	summer sun, part shade	20°F (-7°C)
Daphne caucasica (pinkish white) deciduous daphne	Thymelaeaceae Caucasus	3 ft. (0.9 m)	all summer part shade	hardy
Deutzia pulchra evergreen deutzia	Saxifragaceae Philippines	10 ft. (3 m)	spring sun, part shade	hardy
Diosma ericoides breath of heaven	Rutaceae S. Africa	2 ft. (0.6 m)	winter, spring sun	24°F (-4.5°C)
Erica vagans 'Lyonesse'	Ericaceae Garden	1½ ft. (0.45 m)	summer, autumn shade	hardy
Escallonia X *exoniensis* (pink tinged) The cultivar 'Frades' is dwarf	Saxifragaceae Garden	20 ft. (6 m)	all year sun, part shade	20°F (-7°C)

Euphorbia leucocephala (showy white bracts) flor de pascua	Euphorbiaceae Mexico-C. America	20 ft. (6 m)	midwinter sun	28°F (-2°C)
Exochorda X *macrantha* 'The Bride'	Rosaceae Garden	4 ft. (1.2 m)	late spring sun	hardy
Exochorda racemosa pearly bush	Rosaceae China	15 ft. (4.5 m)	spring sun	hardy
Fabiana imbricata	Solanaceae Peru	8 ft. (2.4 m)	intermittent sun	25°F (-4°C)
Gardenia spp. and cvs. gardenia	Rubiaceae China, Africa	20 ft. (6 m)	summer, winter shade	25°F (-4°C)
Genista monosperma bridal veil broom	Leguminosae Mediterranean	10 ft. (3 m)	late winter, spring sun	27°F (-3°C)
Hamiltonia suaveolens (white to mauve) gidsaw	Rubiaceae India, China	10 ft. (3 m)	autumn, winter light shade	frost tolerant
Helleborus lividus corsicus SS (pale green)	Ranunculaceae Corsica	3 ft. (0.9 m)	late fall, spring shade, part shade	hardy
Helleborus niger SS (white to purple) Christmas rose	Ranunculaceae Europe	1½ ft. (0.45 m)	winter, spring shade, part shade	hardy
Helleborus orientalis SS (greenish or purplish) lenten rose	Ranunculaceae Asia Minor	1½ ft. (0.45 m)	spring shade, part shade	hardy
Lawsonia inermis (creamy white, rose, cinnabar-red) henna	Lythraceae Old World tropics	20 ft. (6 m)	all year sun	40°F (5°C)
Loropetalum chinense VS	Hamamelidaceae China	6 ft. (1.8 m)	spring sun, part shade	27°F (-3°C)
Melastoma candidum (white to reddish)	Melastomataceae Taiwan, SE Asia	10 ft. (3 m)	summer part shade	35°F (2°C)
Murraya paniculata orange jessamine	Rutaceae India, Malaya	12 ft. (3.6 m)	summer-autumn part shade	27°F (-3°C)
Photinia glabra Japanese photinia	Rosaceae Japan	10 ft. (3 m)	summer sun	light frost
Podachaenium eminens	Compositae Mexico, C. America	15 ft. (4.5 m)	spring	25°F (-4°C)
Podalyria calyptrata (mauve, white) sweetpea bush	Leguminosae S. Africa	12 ft. (3.6 m)	spring sun	light frost
Portlandia grandiflora bellflower	Rubiaceae West Indies	18 ft. (5.4 m)	all year sun	45°F (8°C)
Sparmannia africana African linden	Tiliaceae S. Africa	20 ft. (6 m)	midwinter sun	30°F (-1°C)
Tabernaemontana divaricata crape jasmine	Apocynaceae India	15 ft. (4.5 m)	all year sun, part shade	10°F (-12°C)
Turraea obtusifolia SS star bush	Meliaceae S. Africa	4 ft. (1.2 m)	intermittent part shade	26°F (-3°C)

FLOWERS YELLOW TO ORANGE

Abutilon megapotamicum VS (yellow and red) flowering maple	Malvaceae Brazil	10 ft. (3 m)	spring-autumn shade	25°F (-4°C)
Allamanda neriifolia (golden yellow) bush allamanda	Apocynaceae Brazil	3 ft. (0.9 m)	summer sun	35°F (2°C)
Anisacanthus thurberi VS (bright orange) desert honeysuckle	Acanthaceae SW United States, Mexico	5 ft. (1.5 m)	spring-summer sun, dry	hardy
Aphelandra squarrosa (yellow bracts) zebra plant	Acanthaceae Brazil	3 ft. (0.9 m)	all year part shade	35°F (1.5°C)

Asteriscus sericeus	Compositae Canary Is.	3 ft. (0.9 m)	summer sun	25°F (-4°C)
Berberis darwinii (orange, yellow)	Berberidaceae Chile	10 ft. (3 m)	spring sun, part shade	hardy
Berberis X *stenophylla* (golden yellow)	Berberidaceae Garden	6 ft. (1.8 m)	spring sun	hardy
Caesalpinia gilliesii (yellow with red stamens) bird of paradise	Leguminosae S. America	10 ft. (3 m)	summer sun, drainage	27°F (-3°C)
Cestrum aurantiacum (orange) orange cestrum	Solanaceae C. America	8 ft. (2.4 m)	spring, summer sun, part shade	25°F (-4°C)
Corokia cotoneaster	Cornaceae New Zealand	10 ft. (3 m)	spring sun, part shade	10°F (-12°C)
Correa backhousiana (chartreuse)	Rutaceae	5 ft. (1.5 m)	winter sun, part shade	20°F (-7°C)
Crotalaria agatiflora (chartreuse) canary bird bush	Leguminosae Africa	12 ft. (2.4 m)	summer-autumn sun, part shade	28°F (-2°C)
Cytisus canariensis Canary broom	Leguminosae Canary Is.	6 ft. (1.8 m)	spring-summer sun, drainage	15°F (-9°C)
Euryops pectinatus golden shrub daisy	Compositae S. Africa	3 ft. (0.9 m)	all year sun, drainage	light frost
Euryops speciosissimus S. African tree daisy	Compositae S. Africa	6 ft. (1.8 m)	winter sun, drainage	light frost
Forsythia X *intermedia* (many cultivars)	Oleaceae Asia	10 ft. (3 m)	winter-spring sun	hardy
Galphimia glauca shower of gold	Malpighiaceae Mexico-Guatemala	6 ft. (1.8 m)	summer sun	28°F (-2°C)
Gamolepis *chrysanthemoides*	Compositae S. Africa		all year sun	25°F (-4°C)
Gnidia polystachya SS (pale yellow)	Thymelaeaceae S. Africa	6 ft. (1.8 m)	spring sun	30°F (-1°C)
Halimium lasianthum SS	Cistaceae Portugal	3 ft. (0.9 m)	spring sun	24°F (-4.5°C)
Hibbertia cuneiformis VS	Dilleniaceae Australia	6 ft. (1.8 m)	spring sun, drainage	light frost
Holmskioldia sanguinea VS (terracotta) Chinese hat plant	Verbenaceae Himalayas	20 ft. (6 m)	all year sun, part shade	30°F (-1°C)
Kerria japonica 'Pleniflora' VS	Rosaceae China	8 ft. (2.4 m)	spring part shade	hardy
Michelia figo (cream-yellow) banana shrub	Magnoliaceae China	20 ft. (6 m)	spring part shade	25°F (-4°C)
Ochna kirkii Mickey Mouse plant	Ochnaceae E. Africa	15 ft. (4.5 m)	summer part shade	40°F (5°C)
Phlomis fruticosa SS Jerusalem sage	Labiatae S. Europe	4 ft. (1.2 m)	early summer sun	24°F (-4.5°C)
Portlandia domingensis (greenish yellow) Dominica bellflower	Rubiaceae Dominica	12 ft. (3.6 m)	intermittent sun	40°F (5°C)
Potentilla fruticosa SS (many cultivars, colors) cinquefoil	Rosaceae Circumpolar	4 ft. (1.2 m)	summer-autumn sun	hardy
Pultenaea villosa VS (yellow) bush pea	Leguminosae Australia	6 ft. (1.8 m)	spring-summer sun	hardy
Reinwardtia indica SS yellow flax	Linaceae India	4 ft. (1.2 m)	autumn-winter sun, part shade	24°F (-4.5°C)
Rhigozum obovatum karroo gold	Bignoniaceae S. Africa	8 ft. (2.4 m)	summer sun	hardy
Rhododendron sect. Vireya hybrids (orange-yellow)	Ericaceae Garden	5 ft. (1.5 m)	winter part shade	30°F (-1°C)
Sophora chrysophylla (bright yellow) mamane	Leguminosae Hawaii	20 ft. (6 m)	spring-summer sun	28°F (-2°C)

Spartium junceum Spanish broom	Leguminosae Mediterranean	10 ft. (3 m)	spring, summer sun	18°F (-7°C)
Stifftia chrysantha (yellow-orange)	Compositae Brazil	15 ft. (4.5 m)	all year sun	35°F (2°C)
Turnera ulmifolia yellow alder	Turneraceae Trop. America	2 ft. (0.6 m)	all year sun, part shade	30°F (-1°C)

FLOWERS PINK TO RED

Adenium obesum (deep pink) desert rose	Apocynaceae E. Africa	15 ft. (4.5 m)	spring sun	40°F (4.5°C)
Agapetes serpens VS (salmon and red)	Ericaceae W. China, Himalayas	3 ft. (0.9 m)	winter part shade	25°F (-4°C)
Alberta magna (scarlet)	Rubiaceae S. Africa	15 ft. (4.5 m)	winter-spring sun	light frost
Aphelandra sinclairiana (rose-red, orange bracts)	Acanthaceae Costa Rica-Panama	15 ft. (4.5 m)	spring-autumn light shade	35°F (1.5°C)
Begonia coccinea (coral-red)	Begoniaceae Trop. America	10 ft. (3 m)	all year shade	30°F (-1°C)
Begonia corallina (rosy-pink)	Begoniaceae Trop. America	10 ft. (3 m)	all year shade	30°F (-1°C)
Bouvardia leiantha (coral-rose) 'Fire Chief'	Rubiaceae Garden	3 ft. (0.9 m)	spring-autumn sun, part shade	light frost
Bouvardia ternifolia SS (red)	Rubiaceae W. Texas-Mexico	6 ft. (1.8 m)	late summer sun	hardy
Brachysema lanceolatum VS (red) scimitar shrub	Leguminosae W. Australia	5 ft. 1.5 m	spring-autumn sun, part shade	30°F (-1°C)
Brugmansia sanguinea (orange, red)	Solanaceae Peru	15 ft. (4.5 m)	summer sun	32°F (0°C)
Burchellia bubalina (coral red) wild pomegranate	Rubiaceae S. Africa	10 ft. (3 m)	summer shade	light frost
Caesalpinia pulcherrima (red-yellow) pride of Barbados	Leguminosae West Indies	15 ft. (4.5 m)	all year sun	28°F (-2°C)
Cantua buxifolia VS (cerise) flower of Incas	Polemoniaceae Andes	5 ft. (1.5 m)	spring-summer light shade	27°F (-3°C)
Centradenia grandifolia (rose-pink)	Melastomataceae Mexico-Guatemala	5 ft. (1.5 m)	winter light shade	30°F (-1°C)
Cestrum elegans (rosy red)	Solanaceae Mexico	10 ft. (3 m)	all year sun, part shade	25°F (-4°C)
Cestrum newellii crimson or scarlet) red cestrum	Solanaceae Garden	10 ft. (3 m)	spring-summer sun, part shade	25°F (-4°C)
Chorizema cordatum SS (orange and red) flame pea	Leguminosae Australia	5 ft. (1.5 m)	spring-summer sun, part shade	light frost
Chorizema ilicifolium SS (orange and red)	Leguminosae Australia	3 ft. (0.9 m)	spring-summer sun, part shade	light frost
Chorizema varium SS (orange and red)	Leguminosae Australia	3 ft. (0.9 m)	spring-summer sun, part shade	light frost
Clianthus formosus SS VS (scarlet) glory pea	Leguminosae Australia	4 ft. (1.2 m)	summer hot, dry	light frost
Coleonema pulchrum (pink) pink breath of heaven	Rutaceae S. Africa	5 ft. (1.5 m)	autumn-spring sun, drainage	24°F (-4.5°C)
Correa X harrisii (clear red)	Rutaceae Garden	2½ ft. (0.75 m)	winter sun, part shade	20°F (-7°C)
Correa pulchella (pink)	Rutaceae Australia	2½ ft. (0.75 m)	winter sun, part shade	20°F (-7°C)

Cuphea ignea SS (red) cigar plant	Lythraceae Mexico	1 ft. (0.3 m)	summer-autumn sun, part shade	25°F (-4°C)
Cuphea micropetala SS (yellow and red)	Lythraceae Mexico	2 ft. (0.6 m)	summer-autumn sun, part shade	25°F (-4°C)
Cytisus scoparius Dallimore hybrids	Leguminosae Garden	8 ft. (2.4 m)	spring-fall part shade	hardy
Daphne cneorum 'Ruby Glow'	Thymelaeaceae Garden	1 ft. (0.3 m)	spring-fall part shade	hardy
Daphne odora 'Rose Queen'	Thymelaeaceae Garden	4 ft. (1.2 m)	early spring part shade	hardy
Erica herbacea cvs. (white, pink, red)	Ericaceae European Alps	18 ft. (5.4 m)	winter-late spring shade	hardy
Erica mammosa 'Jubilee' (salmon pink)	Ericaceae S. Africa	3 ft. (0.9 m)	spring shade	25°F (-4°C)
Erica vagans 'Mrs. D. F. Maxwell'	Ericaceae Garden	1 ft. (0.9 m)	summer-autumn shade	hardy
Erythrina zeyheri SS (bright red) prickly cardinal	Leguminosae S. Africa	1½ ft. (0.45 m)	spring summer	20°F (-7°C)
Euphorbia milii SS (pink, red, yellow, orange) crown of thorns	Euphorbiaceae Malagasy	4 ft. (1.2 m)	all year part shade, sun	light frost
Euphorbia pulcherrima SS (red, pink, white cvs.) poinsettia	Euphorbiaceae Mexico	8 ft. (2.4 m)	winter sun	25°F (-4°C)
Hakea cucullata (pink)	Proteaceae Australia	14 ft. (4.2 m)	spring sun, drainage	light frost
Hakea multilineata (deep red)	Proteaceae Australia	12 ft. (3.6 m)	spring sun, drainage	light frost
Helianthemum nummularium SS (yellow, peach, pink, red) sunrose	Cistaceae Mediterranean	3 ft. (0.9 m)	spring-early summer sun, drainage	hardy
Holmskiolida sanguinea VS (rust) Chinese hat plant	Verbenaceae Himalayas	15 ft. (4.5 m)	all year sun, part shade	22°F (-5.5°C)
Iochroma coccineum (scarlet)	Solanaceae C. America	8 ft. (2.4 m)	summer-autumn sun, part shade	light frost
Jatropha multifida SS (scarlet) coral plant	Euphorbiaceae Trop. America	20 ft. (6 m)	all year sun, dry	32°F (0°C)
Justicia brandegeana (white with copper-red bracts) shrimp plant	Acanthaceae Mexico	4 ft. (1.2 m)	all year sun, part shade	light frost
Luculia gratissima (pink-rose)	Rubiaceae Himalayas	8 ft. (2.4 m)	autumn-winter part shade	30°F (-1°C)
Malpighia coccigera (rose) dwarf holly	Malpighiaceae West Indies	6 ft. (1.8 m)	summer-autumn sun, part shade	(30°F) (-1°C)
Malpighia glabra (rose to red) Barbados cherry	Malpighiaceae Texas-S. America	10 ft. (3 m)	summer-autumn sun, part shade	28°F (-2°C)
Malvaviscus arboreus mexicanus (red)	Malvaceae Mexico	15 ft. (4.5 m)	summer-autumn shade, part shade	25°F (-4°C)
Medinilla magnifica (watermelon pink) kapa-kapa	Melastomataceae Philippines	10 ft. (3 m)	spring-summer part shade	65°F (18°C)
Megaskepasma erythrochlamys (bracts rose, flowers white, flushed rose) Brazilian red-cloak	Acanthaceae S. America	10 ft. (3 m)	spring-autumn sun, part shade	28°F (-2°C)

Mussaenda erythrophylla (sulphur yellow with scarlet sepal) flag bush	Rubiaceae Trop. Africa	30 ft. (9 m)	spring-summer part shade	40°C (5°C)
Nandina domestica (pinkish flowers, red berries) heavenly bamboo	Berberidaceae China, Japan	8 ft. (2.4 m)	spring-summer sun, part shade	hardy
Nymania capensis (rose red with inflated rosy pods) Chinese lanterns	Meliaceae S. Africa	10 ft. (3 m)	early spring sun	25°F (-4°C)
Paeonia suffruticosa SS (rose, red, white, yellow) tree peony	Paeoniaceae China	6 ft. (2.4 m)	spring sun, part shade	hardy
Pentas lanceolata SS (pink, red, white, lavender)	Rubiaceae Africa	5 ft. (1.5 m)	all year part shade	light frost
Pimelia ferruginea SS (rosy pink) rosy rice flower	Thymelaeaceae Australia	3 ft. (0.9 m)	intermittent part shade	light frost
Rhododendron sect. Vireya hybrids	Ericaceae Garden	5 ft. (1.5 m)	winter part shade	30°F (-1°C)
Rhodomyrtus tomentosa (rose-pink) downy myrtle	Myrtaceae India to China, Philippines	5 ft. (1.5 m)	late spring sun, part shade	30°F (-1°C)
Russellia equisetiformis VS (coral-red) coral-bell bush	Scrophulariaceae Mexico	4 ft. (1.2 m)	all year sun	light frost
Ruttya fruticosa (yellow to scarlet)	Acanthaceae Trop. Africa	12 ft. (3.6 m)	summer sun, part shade	30°F (-1°C)
Salvia fulgens (scarlet) cardinal salvia	Labiatae Mexico	3 ft. (0.9 m)	late spring-summer sun, drainage	tender
Salvia greggii (rosy-red) autumn sage	Labiatae Mexico	4 ft. (1.2 m)	fall sun	25°F (-4°C)
Sanchesia speciosa (yellow with red bracts)	Acanthaceae Garden	6 ft. (1.8 m)	all year sun, shade	30°F (-1°C)
Schotia afra (bright red) Hottentot's bean	Leguminosae S. Africa	10 ft. (3 m)	spring sun	30°F (-1°C)
Stachytarpheta mutabilis (crimson-pink)	Verbenaceae Trop. America	6 ft. (1.8 m)	all year sun	30°F (-1°C)
Sutherlandia frutescens (scarlet)	Leguminosae S. Africa	8 ft. (2.4 m)	winter-spring	28°F (-2°C)
Tamarix parviflora (pink, white)	Tamaricaceae Europe, Asia	12 ft. (3.6 m)	spring sun, dry	20°F (-7°C)
Tecoma garrocha VS (coral or salmon with scarlet tube) Argentine tecoma	Bignoniaceae Argentina	10 ft. (3 m)	summer sun, heat	light frost
Templetonia retusa (brick-red) cockies tongues	Leguminosae Australia	6 ft. (1.8 m)	winter, spring sun, part shade	25°F (-4°C)
Weigelia cvs. (yellow, pink, red)	Caprifoliaceae East Asia	10 ft. (3 m)	spring sun, part shade	hardy
Woodfordia fruticosa (deep rose)	Lythraceae Malagasy, S. Asia	15 ft. (4.5 m)	spring	28°F (-2°C)

FLOWERS BLUE, VIOLET, PURPLE

Alyogyne hakeaefolia (lilac-blue)	Malvaceae Australia	6 ft. (1.8 m)	spring-summer sun	hardy

Caryopteris X *clandonensis* 'Heavenly Blue' blue mist	Verbenaceae Garden	2 ft. (0.6 m)	fall sun	hardy
Caryopteris incana SS (lavender-blue) blue spiraea	Verbenaceae China, Japan	3 ft. (0.9 m)	summer, fall sun	light frost
Clerodendrum ugandense (blue) blue clerodendrum	Verbenaceae Trop. Africa	10 ft. (3 m)	summer sun, part shade	25°F (-4°C)
Cuphea hyssopifolia SS (white, pink, purple) false heather	Lythraceae Mexico	2 ft. (0.6 m)	summer sun, part shade	25°F (-4°C)
Duranta repens (blue flowers, golden berries) sky flower	Verbenaceae Trop. America	15 ft. (4.5 m)	all year sun	25°F (-4°C)
Eranthemum pulchellum (blue)	Acanthaceae Trop. Asia	4 ft. (1.2 m)	spring sun, part shade	32°F (0°C)
Eupatorium sordidum SS (violet blue) mist flower	Compositae Mexico	5 ft. (1.5 m)	summer sun, part shade	20°F (-7°C)
Grewia caffra VS (lavender) lavender star flower	Tiliaceae S. Africa	10 ft. (3 m)	late spring-fall sun	24°F (-4.5°C)
Grewia occidentalis VS (lavender) four corners	Tiliaceae S. Africa	10 ft. (3 m)	all year sun, part shade	28°F (-2°C)
Hebe 'Autumn Glory' (dark lavender-blue)	Scrophulariaceae Garden	2 ft. (0.6 m)	late summer, fall sun, part shade	15°F (-9°C)
Hydrangea macrophylla (blue, pink) lace cap hydrangea	Saxifragaceae China, Japan	12 ft. (3.6 m)	summer, autumn part shade	hardy
Iboza riparia SS (lilac, purple) misty plume bush	Labiatae S. Africa	6 ft. (1.8 m)	autumn, winter shade	light frost
Iochroma cyaneum (blue, purple)	Solanaceae Colombia	8 ft. (2.4 m)	summer, autumn sun, part shade	light frost
Justicia cydoniifolia VS (white flowers, red-violet bracts) Brazilian bower plant	Acanthaceae Brazil	15 ft. (4.5 m)	late spring-early winter part shade	tender
Lavandula dentata (purple) French lavender	Labiatae Mediterranean	3 ft. (0.9 m)	spring-autumn sun, drainage	24°F (-4.5°C)
Leucophyllum frutescens (rosy purple) Texas ranger	Scrophulariaceae Texas, Mexico	10 ft. (3 m)	summer sun, dry	20°F (-7°C)
Mackaya bella (mauve)	Acanthaceae S. Africa	5 ft. (1.5 m)	spring, summer shade, part shade	27°F (-3°C)
Melastoma malabathricum (purple, rosy-mauve) Indian rhododendron	Melastomataceae India, to Queensland	8 ft. (2.4 m)	spring-summer part shade	35°F (1.5°C)
Melastoma sanguineum (purple)	Melastomataceae Malay Peninsula-Java	20 ft. (6 m)	summer part shade	28°F (-2°C)
Nierembergia scoparia SS (white or violet) cup flower	Solanaceae Chile	3 ft. (0.9 m)	summer sun, drainage	20°F (-7°C)
Polygala X *dalmaisiana* (rosy red, purplish) sweetpea shrub	Polygalaceae Garden	5 ft. (1.5 m)	all year sun, part shade	24°F (-4.5°C)
Polygala virgata (purple)	Polygalaceae S. Africa	8 ft. (2.4 m)	autumn-winter sun, drainage	hardy
Prostanthera denticulata (lavender)	Labiatae Australia	6 ft. (1.8 m)	spring sun	light frost
Prostanthera magnifica (blue, red-orange calyx)	Labiatae W. Australia	5 ft. (1.5 m)	spring sun	tender

Prostanthera rotundifolia (lilac blue) mint bush	Labiatae Australia	6 ft. (1.8 m)	spring sun, part shade	light frost
Rosmarinus officinalis VS (lilac to purple) rosemary	Labiatae Europe	3 ft. (0.9 m)	all year sun	hardy
Ruellia macrantha SS (rosy-purple)	Acanthaceae Brazil	6 ft. (1.8 m)	spring, summer part shade	tender
Salvia leucantha (purple or rose) Mexican bush sage	Labiatae Mexico	4 ft. (1.2 m)	summer, autumn sun	25°F (-4°C)
Sollya heterophylla VS (blue) Australian bluebell	Pittosporaceae Australia	3 ft. (0.9 m)	summer sun, part shade	25°F (-4°C)
Stachytarpheta frantzii (purple)	Verbenaceae Costa Rica	6 ft. (1.8 m)	winter, summer	40°F (5°C)
Sutera grandiflora SS (blue) wild phlox	Scrophulariaceae S. Africa	3 ft. (0.9 m)	all year sun, part shade	25°F (-4°C)
Syringa laciniata (lilac)	Oleaceae NW China	6 ft. (1.8 m)	spring sun, part shade	hardy
Syringa vulgaris cvs. (lilac, white) common lilac	Oleaceae SE Europe	20 ft. (6 m)	spring sun, part shade	hardy
Teucrium fruticans (blue) germander	Labiatae Europe	4 ft. (1.2 m)	summer sun	hardy
Thunbergia erecta (blue-purple) king's mantle	Acanthaceae Trop. Africa	6 ft. (1.8 m)	all year sun	35°F (2°C)
Vitex agnus-castus (lilac-blue) chaste tree	Verbenaceae S. Europe	10 ft. (3 m)	late summer-autumn sun	20°F (-7°C)

APPENDIX III
ADDITIONAL VINING PLANTS OF MERIT

Where temperature is not given the plants are usually grown as annuals. See shrub list for other plants that can be used as spillers and sprawlers for low walls, and wall shrubs that can be used as vines if supported. Flowering in tropical areas where day length is almost constant is often associated with the dry season. Height and spread of vines is determined by the support provided.

FLOWERS WHITE

Araujia sericifera	Asclepiadaceae	40 ft.	spring-autumn	25°F
white bladder flower	S. America	(12 m)	sun	(-4°C)
Begonia glabra	Begoniaceae	6 ft.	winter, spring	35°F
climbing begonia	Trop. America	(1.8 m)	shade	(1.5°C)
Camoensia maxima	Leguminosae	30 ft.	summer-autumn	40°F
(white, edged gold)	Trop. Africa	(9 m)	sun	(5°C)
Chonemorpha fragrans	Apocynaceae	30 ft.	summer	27°F
	SE Asia	(9 m)	sun	(-3°C)
Clerodendrum thomsoniae	Verbenaceae	15 ft.	spring to autumn	35°F
(white with red calyx)				
glory bower	Trop. Africa	(4.5 m)	part shade	(1.5°C)
Congea tomentosa	Verbenaceae	30 ft.	all year	28°F
(white to lilac)				
shower orchid	Burma-S. China	(9 m)	sun	(-2°C)
Hylocereus undatus	Cactaceae	40 ft.	summer-autumn	28°F
night-blooming cereus	Trop. America	(12 m)	sun, part shade	(-2°C)
Ipomoea alba	Convolvulaceae	40 ft.	summer-autumn	
moon flower	Tropics	(12 m)	sun	
Philadelphus mexicanus	Saxifragaceae	15 ft.	summer	22°F
mock orange	Mexico	(4.5 m)	sun	(-5.5°C)
Pithecoctenium echinatum	Bignoniaceae	30 ft.	summer	30°F
monkey comb	Trop. America	(9 m)	sun	(-1°C)
Polygonum aubertii	Polygonaceae	30 ft.	summer	20°F
(greenish white)				
silver-lace vine	China	(9 m)	sun	(-7°C)
Porana paniculata	Convolvulaceae	30 ft.	winter	30°F
snow creeper, bridal				
bouquet, Christmas vine	N. India-Burma	(9 m)	part shade	(-1°C)
Porana volubilis	Convolvulaceae	30 ft.	intermittent	35°F
bridal creeper	Burma-Thailand	(9 m)	part shade	(1.5°C)
Selenicereus grandiflorus	Cactaceae	20 ft.	summer	30°F
(white, salmon outside)				
queen of the night	Jamaica, Cuba	(6 m)	sun	(-1°C)
Turbina corymbosa	Convolvulaceae	20 ft.	winter	35°F
pascua	Trop. America	(6 m)	sun	(1.5°C)

FLOWERS YELLOW-ORANGE

Macfadyena dentata	Bignoniaceae	30 ft.	summer	30°F
yellow trumpet vine	Brazil	(9 m)	sun	(-1°C)
Merremia tuberosa	Convolvulaceae	30 ft.	summer-autumn	35°F
wood rose	Trop. America	(9 m)	sun	(1.5°C)
Odontadenia grandiflora	Apocynaceae	30 ft.	intermittent	40°F
(apricot-yellow)	Trop. America	(9 m)	sun	(5°C)
Tristellateia australasiae	Malpighiaceae	30 ft.	all year	40°F
bagnit	SE Asia-New Caledonia	(9 m)	sun	(5°C)
Tropaeolum majus	Tropaeolaceae	15 ft.	summer-autumn	32°F
(yellow, orange, red)				
nasturtium	Mexico, S. America	(4.5 m)	sun, part shade	(0°C)

Tropaeolum speciosum (vermilion, yellow)	Tropaeolaceae Chile	15 ft. (4.5 m)	summer, autumn sun, part shade	32°F (0°C)

FLOWERS PINK TO RED

Asarina erubescens (rose red) maurandia	Scrophulariaceae Mexico	10 ft. (3 m)	spring-autumn sun	32°F (0°C)
Argyreia nervosa (rose) woolly morning glory	Convolvulaceae India	30 ft. (9 m)	spring-summer sun	35°F (1.5°C)
Bauhinia kockiana (scarlet to yellow)	Leguminosae Sumatra-Malaya	15 ft. (4.5 m)	summer-autumn part shade	40°F (5°C)
Bignonia capreolata (orange-red) trumpet flower	Bignoniaceae S. United States	50 ft. (15 m)	spring-summer part shade	10°F (-12°C)
Bomarea caldasii (red and yellow) bomarea	Amaryllidaceae Trop. America	15 ft. (4.5 m)	spring-autumn shade	27°F (-3°C)
Campsis grandiflora (scarlet) Chinese trumpet creeper	Bignoniaceae China	30 ft. (9 m)	summer-autumn sun	-30°F (-35°C)
Campsis radicans (orange, scarlet) trumpet creeper	Bignoniaceae E. United States	50 ft. (15 m)	summer sun	-30°F (-35°C)
Campsis X *tagliabuana* 'Mme. Galen' (salmon red)	Bignoniaceae Garden	30 ft. (9 m)	summer sun	-30°F (-35°C)
Camptosema rubicundum (red)	Leguminosae Argentina	15 ft. (4.5 m)	summer sun	25°F (-4°C)
Clerodendrum splendens (bright scarlet or yellow) climbing scarlet clerodendrum	Verbenaceae Trop. Africa	30 ft. (9 m)	intermittent sun	35°F (1.5°C)
Clianthus puniceus (crimson) parrot's beak	Leguminosae New Zealand	8 ft. (2.4 m)	summer sun, part shade	28°F (-2°C)
Combretum paniculatum (scarlet) burning bush, flame creeper	Combretaceae Trop. Africa	30 ft. (9 m)	midwinter-late spring sun	30°F (-1°C)
Cydista aequinoctialis (lavender to rose) garlic vine	Bignoniaceae Trop. America	30 ft. (9 m)	spring, summer sun	35°F (1.5°C)
Eccremocarpus scaber (orange-red) Chilean glory flower	Bignoniaceae Chile	12 ft. (3.6 m)	summer sun, part shade	25°F (-4°C)
Hidalgoa spp. (orange-red to scarlet)	Compositae C. and S. America	20 ft. (6 m)	winter-spring sun	30°F (-1°C)
Hoya carnosa (dusty pink) wax vine	Asclepiadaceae Australia	15 ft. (4.5 m)	spring-summer part shade	28°F (-2°C)
Ipomoea coccinea (scarlet) scarlet star	Convolvulaceae Arizona, N. Mexico	10 ft. (3 m)	summer-autumn sun, part shade	
Ipomoea horsfalliae (rose-pale purple) princess vine	Convolvulaceae West Indies	30 ft. (4.5 m)	summer-winter sun	35°F (1.5°C)
Ipomoea quamoclit (scarlet) cypress vine	Convolvulaceae Trop. America	20 ft. (6 m)	all year sun, part shade	
Kennedia prostrata (scarlet) coral pea	Leguminosae Australia	prostrate	early summer sun	32°F (0°C)
Kennedia rubicunda (red) coral pea	Leguminosae Australia	prostrate	early summer sun	30°F (-1°C)

Lapageria rosea	Liliaceae	12 ft.	summer-autumn	15°F
(rose-red)				
Chilean bellflower	Chile	(3.6 m)	shade	(-9°C)
Lathyrus odoratus	Leguminosae	8 ft.	spring-summer	
(many colors)				
sweet pea	Italy	(2.4 m)	sun	
Lopezia hirsuta	Onagraceae	prostrate	spring-autumn	30°F
(pink to rose)				
mosquito plant	Mexico		part shade	(-1°C)
Manettia inflata	Rubiaceae	6 ft.	summer	32°F
(scarlet, yellow tipped)				
manettia	Brazil	(1.8 m)	part shade	(0°C)
Marianthus ringens	Pittosporaceae	6 ft.	spring	hardy
(red)				
red-bell-climber	Australia	(1.8 m)	sun	
Mina lobata	Convolvulaceae	20 ft.	summer	25°F
(red turning yellow and white)				
Spanish flag	Mexico	(6 m)	sun	(-4°C)
Mucuna bennettii	Leguminosae	70 ft.	winter-spring	40°F
(orange-red-scarlet)				
New Guinea creeper	New Guinea	(21 m)	sun, part shade	(5°C)
Mutisia spp.	Compositae	20 ft.	summer	25°F
(yellow, pink, red)	Andes, S. America	(6 m)	sun	(-4°C)
Pandorea jasminoides	Bignoniaceae	30 ft.	summer-autumn	30°F
(white, pink)				
bower vine	Australia	(9 m)	sun	(-1°C)
Pereskia aculeata	Cactaceae	30 ft.	intermittent	32°F
(white, pink, yellow)				
lemon vine	Trop. America	(9 m)	sun	(0°C)
Phaseolus coccineus	Leguminosae	20 ft.	summer	32°F
(scarlet)				
scarlet runner bean	Mexico, S. America	(6 m)	sun	(0°C)
Pseudocalymma alliaceum	Bignoniaceae	30 ft.	spring	40°F
(deep rose-pink)				
garlic vine	Mexico-S. America	(9 m)	sun	(5°C)
Quisqualis indica	Combretaceae	30 ft.	spring to autumn	34°F
(white to red)				
Rangoon creeper	SE Asia-New Guinea	(9 m)	sun, part shade	(1°C)
Strongylodon lucidus	Leguminosae	30 ft.	summer	40°F
kaiwi	Hawaiian Islands	(9 m)	part shade	(5°C)
Tecomanthe hillii	Bignoniaceae	30 ft.	intermittent	40°F
(rose)				
pink trumpet vine	NE Australia	(9 m)	part shade	(5°C)

FLOWERS BLUE, VIOLET, PURPLE

Asarina barclaiana	Scrophulariaceae	10 ft.	spring-autumn	32°F
(purple, violet)				
maurandia	Mexico	(3 m)	sun	(0°C)
Clitorea ternatea	Leguminosae	10 ft.	summer	40°F
(bright dark blue)				
butterfly pea	Pantropical	(3 m)	sun, part shade	(5°C)
Cobaea scandens	Polemoniaceae	20 ft.	summer-autumn	24°F
(violet, white)				
cup and saucer vine	Mexico	(6 m)	sun	(-4.5°C)
Cryptostegia grandiflora	Asclepiadaceae	12 ft.	summer	26°F
(lilac purple)				
rubber vine	Africa	(3.6 m)	sun	(-3°C)
Cymbalaria muralis	Scrophulariaceae	3 ft.	spring-autumn	
(lilac-blue)				
Kenilworth ivy	Europe	(0.9 m)	shade	
Dolichos lablab	Leguminosae	25 ft.	summer	30°F
(purple or white)				
hyacinth bean	Tropics	(7.5 m)	sun	(-1°C)

Dolichos lignosus	Leguminosae	20 ft.	summer	24°F
(white or violet)				
Australian pea vine	Australia	(6 m)	sun	(-4.5°C)
Heterocentron elegans	Melastomataceae	prostrate	summer	27°F
(purple or violet)				
Spanish shawl	Mexico, Guatemala		part shade	(-3°C)
Ipomoea acuminata	Convolvulaceae	30 ft.	spring-autumn	
(bright blue)				
blue dawn flower	Trop. America	(9 m)	sun	
Jacquemontia pentantha	Convolvulaceae	10 ft.	spring	30°F
(lilac-blue)	Florida-S. America	(3 m)	sun, part shade	(-1°C)
Porana grandiflora	Convolvulaceae	30 ft.	late summer-autumn	
(purple)	Nepal-Sikkim	(9 m)	part shade	
Solanum seaforthianum	Solanaceae	20 ft.	summer	30°F
(blue-purple)				
Brazilian nightshade	Trop. America	(6 m)	partial shade	(-1°C)
Strongylodon macrobotrys	Leguminosae	60 ft.	spring-summer	40°F
(bluish-green)				
jade vine	Philippines	(18 m)	partial shade	(5°C)
Thunbergia laurifolia	Acanthaceae	30 ft.	intermittent	30°F
(light blue-lavender)	Trop. SE Asia	(9 m)	sun, part shade	(-1°C)

APPENDIX IV
ANNUALS AND PERENNIALS USEFUL FOR GROUND COVERS

Bellis perennis	Compositae	6 in.	winter-spring
English daisy	Europe	(15 cm)	several colors
Canna cultivars	Cannaceae	6 ft.	summer
canna	Tropics	(1.8 m)	various colors
Chrysanthemum parthenium	Compositae	3 ft.	summer
'Aureum'			
golden feather	Eurasia	(0.9 m)	white and yellow
Coreopsis pubescens	Compositae	4 ft.	summer
dwarf coreopsis	E. United States	(1.2 m)	yellow
Dianthus deltoides	Caryophyllaceae	12 in.	spring-summer
maiden pink	Eurasia	(30 cm)	pink
Dianthus plumarius	Caryophyllaceae	18 in.	summer-autumn
cottage pink	Eurasia	(45 cm)	various colors
Dyssodia tenuiloba	Compositae	12 in.	summer-autumn
golden fleece	Mexico	(30 cm)	yellow
Iberis sempervirens	Cruciferae	18 in.	winter-spring
evergreen candytuft	Eurasia	(45 cm)	white
Lobelia erinus	Lobeliaceae	6 in.	summer-autumn
annual lobelia	S. Africa	(15 cm)	blue-purple
Lobularia maritima	Cruciferae	6 in.	all year
sweet alyssum	Europe	(15 cm)	several colors
Malcolmia maritima	Cruciferae	15 in.	winter-spring
Virginia stock	Europe	(37.5 cm)	several colors
Phlox drummondii	Polemoniaceae	18 in.	summer
dwarf annual phlox	Texas	(45 cm)	several colors
Portulaca grandiflora	Portulacaceae	6 in.	summer
portulaca, moss rose	Brazil	(15 cm)	many colors
Salvia splendens	Labiatae	3 ft.	summer
scarlet sage	Brazil	(0.9 m)	scarlet
Sanvitalia procumbens	Compositae	6 in.	summer
	Mexico	(15 cm)	yellow and purple
Tagetes patula	Compositae	18 in.	summer
French marigold	Mexico	(45 cm)	yellow-brown
Verbena X *hybrida*	Verbenaceae	12 in.	summer
garden verbena	Garden	(30 cm)	various colors
Viola cornuta	Violaceae	8 in.	winter-spring
viola	S. Europe	(20 cm)	various colors
Viola odorata	Violaceae	6 in.	spring
sweet violet	Eurasia-Africa	(15 cm)	various colors
Viola X *wittrockiana*	Violaceae	8 in.	winter-spring
pansy	Garden	(20 cm)	various colors
Zinnia, dwarf types	Compositae	12 in.	summer
zinnia	Mexico	(30 cm)	many colors

APPENDIX V
ADDITIONAL PLANTS FOR COLOR ON THE GROUND

Many plants may be massed for color on the ground. In certain parts of the tropics bromeliads are used in this fashion, providing color from both foliage and flower. Orchids, such as *Cymbidium* and *Epidendrum* species, may also be used where the climate permits. Some of the sprawling shrubs and vines are suitable as ground covers.

Achillea tomentosa	Compositae	10 in.	summer	0°F
woolly yarrow	Eurasia	(25 cm)	yellow	(-18°C)
Amaryllis belladonna	Amaryllidaceae	3 ft.	autumn	25°F
naked lady	S. Africa	(0.9 m)	pink	(-4°C)
Anthurium X *cultorum*	Araceae	3 ft.	all year	40°F
anthurium	Garden	(0.9 m)	several colors	(4.5°C)
Armeria maritima	Plumbaginaceae	12 in.	all year	0°F
sea pink	Europe	(30 cm)	pink	(-18°C)
Begonia X *semperflorens-cultorum*	Begoniaceae	18 in.	summer-autumn	35°F
bedding begonia	Garden	(45 cm)	white-rose	(2°C)
Bergenia crassifolia	Saxifragaceae	20 in.	winter	15°F
bergenia	Himalayas	(50 cm)	rose-lilac	(-9°C)
Campanula elatines garganica	Campanulaceae	6 in.	summer	0°F
bellflower	S. Europe	(15 cm)	violet	(-18°C)
Campanula poscharskyana	Campanulaceae	12 in.	spring-summer	0°F
Serbian bellflower	Yugoslavia	(30 cm)	blue	(-18°C)
Carissa macrocarpa dwarf cultivars	Apocynaceae	2 ft.	spring-autumn	26°F
carissa	Africa	(60 cm)	white	(-3°C)
Centaurea cineraria	Compositae	12 in.	summer	20°F
dusty miller	Europe	(30 cm)	purple	(-7°C)
Ceratostigma griffithii	Plumbaginaceae	3 ft.	summer	27°F
Burmese plumbago	India	(0.9 m)	blue	(-3°C)
Ceratostigma plumbaginoides	Plumbaginaceae	12 in.	summer	15°F
dwarf plumbago	China	(30 cm)	bluish	(-9°C)
Ceratostigma willmottianum	Plumbaginaceae	4 ft.	summer-autumn	15°F
Chinese plumbago	China	(1.2 m)	blue	(-9°C)
Cistus hybrids	Cistaceae	5 ft.	summer	20°F
rockrose	Mediterranean	(1.5 m)	various colors	(-7°C)
Cotoneaster, several low shrubby species	Rosaceae	3 ft.	summer-winter	15°F
cotoneaster	Asia	(0.9 m)	white flowers, red fruit	(-9°C)
Crinum spp.	Amaryllidaceae	4 ft.	intermittent	28°F
spider lily	Tropics, Subtropics	(1.2 m)	white-pink	(-2°C)
Crossandra infundibuliformis	Acanthaceae	3 ft.	summer	30°F
firecracker flower	S. India-Ceylon	(0.9 m)	orange-pink	(-1°C)
Episcia spp.	Gesneriaceae	6 in.	intermittent	35°F
carpet plant	Trop. America	(15 cm)	various colors	(2°C)
Erigeron karvinskianus	Compositae	20 in.	summer-autumn	15°F
vittadinia	Mexico-Panama	(50 cm)	white	(-9°C)
Fittonia verschaffeltii	Acanthaceae	6 in.	summer	35°F
mosaic plant	Colombia-Peru	(15 cm)	colored foliage	(13°C)
Helianthemum nummularium	Cistaceae	8 in.	spring-summer	15°F
sunrose	Mediterranean	(20 cm)	various colors	(-9°C)
Hemerocallis cultivars	Liliaceae	6 ft.	all year	10°F
daylily	Asia	(1.8 m)	yellow, bronze-red	(-12°C)

Hippeastrum cultivars	Amaryllidaceae	2.5 ft.	summer-autumn	25°F
amaryllis	Trop. America	(75 cm)	various colors	(-4°C)
Impatiens repens	Balsaminaceae	6 in.	summer	40°F
creeping impatiens	India-Ceylon	(15 cm)	yellow	(5°C)
Iris cultivars	Iridaceae	4 ft.	most of year	0°F
iris	Worldwide	(1.2 m)	many colors	(-18°C)
Lotus berthelotii	Leguminosae	2 ft.	summer	27°F
parrot's beak	Canary Is.	(60 cm)	scarlet	(-3°C)
Lotus corniculatus	Leguminosae	6 in.	summer-autumn	10°F
birdsfoot trefoil	Eurasia	(15 cm)	yellowish	(-12°C)
Lotus mascaensis	Leguminosae	2 ft.	summer	25°F
	Canary Is.	(60 cm)	yellow	(-4°C)
Lysimachia nummularia	Primulaceae	3 in.	summer	0°F
moneywort	Europe	(7.5 cm)	yellow	(-18°C)
Mazus reptans	Scrophulariaceae	2 in.	spring-summer	0°F
	Himalayas	(5 cm)	blue	(-18°C)
Myoporum parvifolium	Myoporaceae	3 in.	summer	25°F
	Australia	(7.5 cm)	white	(-4°C)
Oenothera speciosa	Onagraceae	12 in.	summer	20°F
Mexican evening primrose	Mexico, Texas	(30 cm)	pink	(-7°C)
Oxypetalum caeruleum	Asclepiadaceae	2 ft.	summer	27°F
	Argentina	(60 cm)	blue	(-3°C)
Phyla nodiflora	Verbenaceae	3 in.	spring-autumn	20°F
lippia	Tropics	(7.5 cm)	lilac-rose	(-7°C)
Plumbago auriculata	Plumbaginaceae	10 ft.	all year	25°F
cape plumbago	S. Africa	(3 m)	white-blue	(-4°C)
Ranunculus repens	Ranunculaceae	2 ft.	spring	hardy
creeping buttercup	N. Temperate	(60 cm)	yellow	
Rosa spp. and cultivars	Rosaceae	2 ft.	all year	15°F
climbing roses	China	(60 cm)	various colors	(-9°C)
Rosmarinus officinalis	Labiatae	2 ft.	winter-spring	10°F
'Prostratus'				
creeping rosemary	Mediterranean	(60 cm)	blue	(-12°C)
Santolina chamaecyparissus	Compositae	2 ft.	summer	20°F
lavender cotton	Mediterranean	(60 cm)	yellow	(-7°C)
Santolina virens	Compositae	2 ft.	summer	20°F
green santolina	Europe	(60 cm)	chartreuse	(-7°C)
Saxifraga stolonifera	Saxifragaceae	2 ft.	spring	tender
strawberry geranium	E. Asia	(50 cm)	light pink	
Spathiphyllum spp.	Araceae	5 ft.	all year	40°F
spathe flower	Trop. America	(1.5 m)	white bracts	(5°C)
Teucrium chamaedrys	Labiatae	6 in.	summer	hardy
'Prostratum'				
prostrate germander	Europe	(15 cm)	lavender	
Trachelospermum	Apocynaceae	2 ft.	summer	20°F
jasminoides				
star jasmine	S. China	(60 cm)	white	(-7°C)
Verbena peruviana cultivars	Verbenaceae	18 in.	summer	27°F
	S. America	(45 cm)	various colors	(-3°C)
Verbena tenera	Verbenaceae	18 in.	summer	20°F
sand verbena	Brazil	(45 cm)	various colors	(-7°C)
Vinca major	Apocynaceae	12 in.	all year	0°F
periwinkle	Europe	(30 cm)	lilac-blue	(-18°C)
Vinca minor	Apocynaceae	12 in.	all year	0°F
periwinkle	Europe	(30 cm)	blue-lilac-blue	(-18°C)

APPENDIX VI

ADDITIONAL COLORFUL CALIFORNIA NATIVE PLANTS OF MERIT

Abronia villosa	Nyctaginaceae	1½ ft.	winter-summer
desert verbena	herb	(45 cm)	purplish-rose
Acacia greggii	Leguminosae	6 ft.	spring
sweet acacia	shrub	(1.8 m)	yellow
Antirrhinum nuttallianum	Scrophulariaceae	3 ft.	spring-winter
violet snapdragon	herb	(0.9 m)	violet
Aquilegia eximia	Ranunculaceae	3 ft.	spring-summer
Van Houtte's columbine	herb	(0.9 m)	scarlet
Arabis blepharophylla	Cruciferae	8 in.	winter-spring
coast rock cress	herb	(20 cm)	rose-purple
Arbutus menziesii	Ericaceae	70 ft.	winter-spring
madrone	tree	(21 m)	white to pink
Armeria maritima	Plumbaginaceae	1½ ft.	spring
sea thrift	herb	(45 cm)	pink
Aster greatai	Compositae	4 ft.	summer-autumn
Greata's aster	herb	(1.2 m)	light purple
Bloomeria crocea	Amaryllidaceae	2 ft.	spring
golden stars	herb	(60 cm)	orange-yellow
Brodiaea elegans	Amaryllidaceae	1½ ft.	spring-summer
harvest brodiaea	herb	(45 cm)	purple
Brodiaea lutea	Amaryllidaceae	2½ ft.	spring-autumn
golden brodiaea	herb	(75 cm)	golden-yellow
Calliandra eriophylla	Leguminosae	3 ft.	winter-spring
fairy duster	shrub	(0.9 m)	rose
Cassia armata	Leguminosae	3½ ft.	spring
senna	shrub	(1 m)	yellow
Collinsia heterophylla	Scrophulariaceae	1½ ft.	spring
Chinese houses	annual herb	(45 cm)	rose-purple
Comarostaphylis diversifolia	Ericaceae	15 ft.	spring
summer holly	shrub	(4.5 m)	white
Cornus nuttallii	Cornaceae	75 ft.	spring-summer
mountain dogwood	tree	(22.5 m)	white
Crossosoma californicum	Crossosomataceae	6 ft.	winter-spring
crabapple bush	shrub	(1.8 m)	white
Delphinium cardinale	Ranunculaceae	6 ft.	spring-summer
scarlet larkspur	herb	(1.8 m)	scarlet
Delphinium parryi	Ranunculaceae	3 ft.	spring
Parry's larkspur	herb	(0.9 m)	purplish-blue
Dicentra formosa	Fumariaceae	1½ ft.	spring-summer
bleeding heart	herb	(45 cm)	rose-purple
Dichelostemma ida-maia	Amaryllidaceae	3 ft.	spring-summer
firecracker flower	herb	(0.9 m)	bright red
Enceliopsis covillei	Compositae	4 ft.	spring
Panamint daisy	herb	(1.2 m)	yellow
Erigeron glaucus	Compositae	1 ft.	spring-summer
seaside daisy	herb	(30 cm)	violet-lavender
Eriogonum arborescens	Polygonaceae	5 ft.	summer
island buckwheat	shrub	(1.5 m)	white
Eriogonum crocatum	Polygonaceae	1 ft.	spring-summer
Conejo buckwheat	herb	(30 cm)	sulfur-yellow
Eriogonum giganteum	Polygonaceae	6 ft.	summer
St. Catherine's lace	shrub	(1.8 m)	white
Galvezia speciosa	Scrophulariaceae	3 ft.	spring
island snapdragon	vining shrub	(0.9 m)	red
Gilia tricolor	Polemoniaceae	1½ ft.	spring
bird's eye gilia	annual herb	(45 cm)	yellow, blue-violet
Holodiscus discolor	Rosaceae	18 ft.	spring-summer
cream bush	shrub	(5.4 m)	creamy-white
Hypericum formosum scouleri	Hypericaceae	2½ ft.	summer
St. John's wort	herb	(75 cm)	yellow

Iris, several species and hybrids	Iridaceae	3 ft.	spring
iris	herb	(0.9 m)	various colors
Isomeris arborea	Capparaceae	5 ft.	most of year
bladder pod	shrub	(1.5 m)	yellow
Justicia californica	Acanthaceae	5 ft.	spring-summer
chuparosa	shrub	(1.5 m)	scarlet-red
Lepechinia calycina	Labiatae	4 ft.	spring
pitcher sage	shrub	(1.2 m)	pinkish
Lilium humboldtii	Liliaceae	6 ft.	summer
Humboldt's lily	herb	(1.8 m)	orange-yellow
Lobelia cardinalis	Lobeliaceae	3½ ft.	summer-autumn
scarlet lobelia	herb	(1 m)	bright red
Lonicera involucrata	Caprifoliaceae	10 ft.	summer
twinberry	shrub	(3 m)	yellow
Mentzelia lindleyi	Loasaceae	2 ft.	spring
blazing star	annual herb	(60 cm)	golden-yellow
Mimulus cardinalis	Scrophulariaceae	2½ ft.	spring-autumn
scarlet monkeyflower	herb	(75 cm)	scarlet
Monolopia lanceolata	Compositae	2 ft.	spring
hilltop daisy	annual herb	(60 cm)	bright yellow
Phacelia campanularia	Hydrophyllaceae	2 ft.	spring
bell-flowered phacelia	annual herb	(60 cm)	deep blue
Philadelphus lewisii gordonianus	Saxifragaceae	10 ft.	spring-summer
Gordon's syringa	shrub	(3 m)	white
Ribes aureum	Saxifragaceae	6 ft.	spring
golden currant	shrub	(1.8 m)	yellow
Ribes malvaceum	Saxifragaceae	6 ft.	winter-spring
pink flowering currant	shrub	(1.8 m)	rose
Ribes sanguineum	Saxifragaceae	9 ft.	spring
fuchsia-flowered gooseberry	shrub	(2.7 m)	red
Rosa nutkana	Rosaceae	6 ft.	spring-summer
Nootka rose	shrub	(1.8 m)	rose-pink
Salvia clevelandii	Labiatae	3 ft.	spring-summer
blue sage	shrub	(0.9 m)	blue-purple
Salvia spathacea	Labiatae	3 ft.	spring
crimson sage	herb	(0.9 m)	purplish-red
Sisyrinchium bellum	Iridaceae	1½ ft.	spring
blue-eyed grass	herb	(45 cm)	violet to blue
Sphaeralcea ambigua	Malvaceae	3½ ft.	spring
desert mallow	herb or shrub	(1 m)	apricot
Stanleya pinnata	Cruciferae	5 ft.	spring-summer
golden prince's plume	herb	(1.5 m)	yellow
Styrax officinalis	Styracaceae	12 ft.	spring
snowdrop bush	shrub	(3.6 m)	white

APPENDIX VII
SOURCES OF PLANTS

Not all plants described or listed in this book are available at all nurseries. Many commercial nurseries must of necessity stock those plants that are well known and provide income from rapid sales. Most nurserymen, however, upon request are willing to obtain the more unusual plants from wholesalers. Again it is essential that the plant is ordered under the correct botanical name.

Botanical gardens and arboreta are continually introducing many new flowering plants by making stock available to nurseries or by selling them at their sales shops or at special annual sales. Local horticultural societies also are often a fine source of unusual material and are interested in promoting the introduction and testing of promising new plants for our gardens.

WHERE TO SEE MATURE SPECIMENS

Specimens of the plants illustrated here may be seen in many collections, particularly in botanical gardens and arboreta. No one garden will display all of them. Notable private gardens are occasionally open to visitors. In some areas certain of the plants illustrated may be used as street and avenue trees and others may be seen in public parks.

Tropical and subtropical flowering plants may be seen in many public gardens throughout the world, although in temperate areas they may be grown only in conservatories. City parks and botanical gardens and arboreta associated with universities and colleges may have fine collections open to the public. Following is a selection from the many public gardens where one may see some of the plants mentioned in this book growing outdoors.

AUSTRALIA

Adelaide Botanic Garden, South Australia
Botanic Garden, Brisbane, Queensland
Flecker Botanic Gardens, Cairns, Queensland
Canberra Botanic Gardens, Australian Capital Territory
Royal Botanic Gardens, Melbourne, Victoria
Kings Park and Botanic Garden, West Perth, Western Australia
Royal Botanic Gardens, Sydney, New South Wales

BRAZIL

Jardim Botânico do Rio de Janeiro
Jardim Botânico de Sao Paulo

COSTA RICA

Las Cruces Tropical Botanical Garden, San Vito

CUBA

Jardín Botánico de Cienfuegos

FIJI ISLANDS

Suva Gardens

HONG KONG

Botanic Gardens

INDIA

Indian Botanic Garden, Calcutta
National Botanic Garden, Lucknow

INDONESIA

Kebun Raya Indonesia, Bogor

JAMAICA

Royal Botanic Gardens (Hope), Kingston

SINGAPORE

Botanic Gardens

SOUTH AFRICA

National Botanic Gardens of South Africa, Kirstenbosch

SRI LANKA

Royal Botanic Gardens, Peradeniya

TAIWAN

Heng-chun Tropical Botanical Garden, Ping-tung

UNITED STATES

Los Angeles State and County Arboretum, Arcadia, California
Rancho Santa Ana Botanic Garden, Claremont, California
Waimea Arboretum, Haleiwa, Oahu, Hawaii
Foster Botanic Garden, Honolulu, Oahu, Hawaii
Pacific Tropical Botanical Garden, Lawai, Kauai, Hawaii
Fairchild Tropical Garden, Miami, Florida
South Coast Botanic Garden, Palos Verdes Peninsula, California
Desert Botanical Garden, Phoenix, Arizona
Santa Barbara Botanic Garden, Santa Barbara, California
The Marie Selby Botanical Garden, Sarasota, Florida
Huntington Botanical Gardens, San Marino, California

USSR

State Nikita Botanical Garden, Yalta

SELECTED BIBLIOGRAPHY

Bailey Hortorium. 1976. *Hortus Third.* 1290 pp. New York: Macmillan.

Barrett, Mary F. 1956. *Common Exotic Trees of South Florida.* 414 pp. Gainesville: University of Florida Press.

Beittel, Will. 1972. *Santa Barbara's Street and Park Trees.* 94 pp. Santa Barbara: Santa Barbara County Horticultural Society. Illustrated with black and white and color photographs.

Blatter, E., and W. S. Millard. 1954. *Some Beautiful Indian Trees.* 2d ed. Rev. by W. T. Stearn. 165 pp. Bombay: Bombay Natural History Society. Illustrated.

Bor, N. L., and M. B. Raizada. 1954. *Some Beautiful Indian Climbers and Shrubs.* 286 pp. Bombay: Bombay Natural History Society. Illustrated with line drawings, black and white photographs, watercolors.

Bruggeman, L. 1957. *Tropical Plants and Their Cultivation.* 228 pp. London: Thames & Hudson. Illustrated with watercolors.

Clay, Horace F., and James C. Hubbard. 1977*a. The Hawai'i Garden. Tropical Exotics.* 266 pp. Honolulu: University Press of Hawaii. Illustrated with color photographs.

———. 1977*b. The Hawai'i Garden. Tropical Shrubs.* 295 pp. Honolulu: University Press of Hawaii. Illustrated with color photographs.

Compton, Robert Harold. 1965. *Kirstenbosch.* 168 pp. Cape Town: Tafelberg-Uitgeevers. Black and white and color photographs.

Cornell, Ralph D. 1978. *Conspicuous California Plants.* Rev. ed. 232 pp. Los Angeles: The Plantin Press. Black and white photographs.

Cowen, D. V. 1965. *Flowering Trees and Shrubs in India.* 4th ed. 142 pp. Bombay: Thacker & Co. Illustrated with watercolors.

DuCane, Florence, and Ella DuCane. 1909. *The Flowers and Gardens of Madeira.* 150 pp. London: Adam & Charles Black. Illustrated with watercolors.

Eliovson, Sima. 1955. *South African Flowers for the Garden.* 306 pp. Cape Town: Howard Timmins. Illustrated with black and white and color photographs.

———. 1962. *Flowering Shrubs, Trees and Climbers for Southern Africa.* 216 pp. Cape Town: Howard Timmins. Illustrated with color photographs.

Elliott, W. Rodger, and David L. Jones. 1980. *Encyclopaedia of Australian Plants.* 336 pp. Melbourne: Lothian Publishing Company. Illustrated with photographs and line drawings. Volume 1 of a series.

Fairall, A. R. 1970. *West Australian Native Plants in Cultivation.* 253 pp. Australia: Pergamon Press. Color photographs.

Gough, Kathleen. 1928. *A Garden Book for Malaya.* 422 pp. London: H. F. & G. Witherby. Illustrated.

Graf, Alfred Byrd. 1974*a. Exotica 3.* 1831 pp. East Rutherford, N.J.: Roehrs Company. Profusely illustrated with black and white and color photographs.

———. 1974*b. Exotic Plant Manual.* 4th ed. 840 pp. East Rutherford, N.J.: Roehrs Company. Illustrated with black and white and color photographs.

———. 1978. *Tropica: Color encyclopedia of exotic plants and trees from the tropics and subtropics.* 1120 pp. Rutherford, N.J.: Roehrs Company. Illustrated with color photographs.

Greene, Wilhelmina F., and Hugo L. Blomquist. 1953. *Flowers of the South.* 208 pp. Chapel Hill: University of North Carolina Press. Illustrated with black and white drawings and watercolors.

Hannau, Hans W., and Jeanne Garrard. *Flowers of the Bahamas.* 64 pp. New York: Hastings House. Illustrated with color photographs.

Hargreaves, Dorothy and Bob Hargreaves. 1970*a. Tropical Trees of the Pacific.* 64 pp. Kailua, Hawaii: Hargreaves Company. Illustrated with color photographs.

———. 1970*b*. *Tropical Blossoms of the Pacific.* 64 pp. Kailua, Hawaii: Hargreaves Company. Illustrated with color photographs.
Harris, Thistle Y. 1953. *Australian Plants for the Garden.* 354 pp. Sydney: Angus & Robertson. Illustrated with black and white photographs.
Harrison, Richmond E. 1974. *Trees and Shrubs.* 408 pp. Wellington: A. H. & A. W. Reed. Illustrated with black and white photographs.
Hastings, George T. 1976. *Trees of Santa Monica.* Rev. ed. by Grace L. Heintz. 213 pp. Santa Monica: Friends of Santa Monica Library. Illustrated with line drawings and black and white photographs.
Herbert, D. A. 1952. *Gardening in Warm Climates.* 245 pp. Sydney: Angus & Robertson. Illustrated with black and white photographs.
Herklots, Geoffrey. 1976. *Flowering Tropical Climbers.* 194 pp. Folkestone, England: Dawson. Illustrated with black and white drawings and watercolors.
Holttum, R. E. 1953. *Gardening in the Lowlands of Malaya.* 323 pp. Singapore: Strait Times Press. Illustrated with line drawings, black and white and color photographs.
Hong Kong Trees. 1975. 107 pp. Hong Kong: Urban Services Department, Government Printer. Illustrated with color photographs.
Hong Kong Shrubs. 1976. 113 pp. Hong Kong: Urban Services Department, Government Printer. Illustrated with color photographs.
Hoyt, Roland Stewart. 1978. *Ornamental Plants for Subtropical Regions.* 485 pp. Anaheim, Calif.: Livingston Press.
Hyams, Edward, and William MacQuitty. 1969. *Great Botanical Gardens of the World.* 288 pp. New York: Macmillan. Black and white and color photographs.
Jackson, Robert. 1953. *Beautiful Gardens of the World.* 88 pp. London: Evans Brothers. Illustrated with black and white photographs.
Jex-Blake, A. J., ed. 1950. *Gardening in East Africa.* 3d ed. 398 pp. London: Longmans, Green & Co. Illustrated with watercolors.
Kuck, Loraine E., and Richard C. Tongg. 1955. *The Modern Tropical Garden.* 250 pp. Honolulu: Tongg Publishing Company. Illustrated with black and white photographs.
———. 1960. *Hawaiian Flowers and Flowering Trees.* 158 pp. Rutland, Vt.: Charles E. Tuttle Co. Illustrated with watercolors.
Kunkel, Günther. 1978. *Flowering Trees in Subtropical Gardens.* 346 pp. The Hague: W. Junk. Illustrated with black and white drawings.
Labadie, Emile L. 1978. *Native Plants for Use in the California Landscape.* 244 pp. Sierra City, Calif.: Sierra City Press. Illustrated with line drawings.
———. 1980. *Ornamental Shrubs for Use in the Western Landscape.* 308 pp. Sierra City, Calif.: Sierra City Press. Illustrated with black and white drawings.
Lenz, Lee W. 1956. *Native Plants for California Gardens.* 166 pp. Claremont: Rancho Santa Ana Botanic Garden. Illustrated with black and white photographs.
Lenz, Lee W., and John Dourley. 1981. *California Native Trees & Shrubs.* 232 pp. Illustrated with line drawings and black and white and color photographs.
Lord, Ernest E. 1956. *Shrubs and Trees for Australian Gardens.* 3d rev. ed. 443 pp. Melbourne: Lothian Publishing Company. Illustrated with black and white and color photographs.
Macmillan, H. F. 1946. *Tropical Planting and Gardening* (with special reference to Ceylon). 560 pp. London: Macmillan. Illustrated with black and white photographs.
Macoby, Stirling. 1979. *Trees for Warm and Temperate Climates.* 182 pp. Sydney: Ure Smith. Illustrated with color photographs.
Martineau, Mrs. Philip. 1924. *Gardening in Sunny Lands* (the Riviera, California, Australia). 296 pp. London: Richard Cobden-Sanderson. Illustrated with black and white photographs.
Matschat, Cecile Hulse. 1935. *Mexican Plants for American Gardens.* 269 pp.

Boston: Houghton Mifflin Company. Illustrated with line drawings and black and white photographs.
Menninger, Edwin A. 1962. *Flowering Trees of the World.* 336 pp. New York: Hearthside Press. Illustrated with color photographs.
———. 1964. *Seaside Plants of the World.* 303 pp. New York: Hearthside Press. Illustrated with black and white photographs.
———. 1970. *Flowering Vines of the World.* 410 pp. New York: Hearthside Press. Illustrated with black and white and color photographs.
Morton, Julia F. 1971. *Exotic Plants.* 160 pp. New York: Golden Press. Illustrated with watercolors.
Muller, Katherine K., Richard E. Broder, and Will Beittel. 1974. *Trees of Santa Barbara.* 248 pp. Santa Barbara: Santa Barbara Botanic Garden. Illustrated with line drawings, black and white and color photographs.
Neal, Marie C. 1965. *In Gardens of Hawaii.* 924 pp. Honolulu: Bishop Museum. Illustrated with line drawings.
O'Gorman, Helen. 1961. *Mexican Flowering Trees and Plants.* 218 pp. Mexico City: Ammex Associados. Illustrated with watercolors.
Pal, B. P., and S. Krishnamurthi. 1967. *Flowering Shrubs.* 155 pp. New Delhi: Indian Council of Agricultural Research. Illustrated with color photographs.
Perry, Bob. 1981. *Trees and Shrubs for Dry California Landscapes.* 184 p. San Demas, Calif. Land Design Publishing. Illustrated with color photographs.
Pertchik, Bernard, and Harriet Pertchik. 1951. *Flowering Trees of the Caribbean.* 125 pp. New York: Rinehart & Co. Illustrated with watercolors.
Rowell, Raymond J. 1980*a*. *Ornamental Flowering Shrubs in Australia.* 264 pp. Sydney: Reed. Illustrated with color photographs.
———. 1980*b*. *Ornamental Flowering Trees in Australia.* 240 pp. Sydney: Reed. Illustrated with color photographs.
Schmidt, Marjorie G. 1980. *Growing California Native Plants.* 366 pp. Berkeley, Los Angeles, London: University of California Press. Illustrated with line drawings and color photographs.
[Singapore]. *Selected Plants and Planting for a Garden City: Forty Popular Climbers.* 95 pp. Singapore: Ministry of Law & National Development. Illustrated with color photographs.
[Singapore]. *Selected Plants and Planting for a Garden City: Forty Shrubs.* 87 pp. Singapore: Ministry of Law and National Development. Illustrated with color photographs.
[Singapore]. *Selected Plants and Planting for a Greener Singapore.* 55 pp. Singapore: Ministry of Law and National Development. Illustrated with color photographs.
Southern Living. Gardening. 1980. *Trees and Shrubs.* 260 pp. Birmingham, Ala.: Oxmoor House. Illustrated.
Storer, Dorothy P. 1958. *Familiar Trees and Cultivated Plants of Jamaica.* 81 pp. London: Macmillan. Illustrated with line drawings.
Stout, Mary, and Madeline Agar. 1921. *A Book of Gardening for the Sub-Tropics.* 200 pp. London: H. F. & G. Witherby. Focused on Cairo.
Sturrock, David and Edwin A. Menninger. 1946. *Shade and Ornamental Trees for South Florida and Cuba.* 172 pp. Stuart, Fla.: Stuart Daily News. Illustrated with black and white photographs.
Sunset Magazine. 1979. *New Western Garden Book.* 512 pp. Menlo Park, Calif.: Lane Publishing. Illustrated.
Tanner, Howard. 1976. *The Great Gardens of Australia.* 198 pp. Melbourne: Macmillan. Black and white photographs.
Thrower, S. L. 1976. *Hong Kong Herbs and Vines.* Rev. ed. 114 pp. Hong Kong: Government Printer. Illustrated with color photographs.
Van der Spuy, Una. 1954. *Ornamental Shrubs and Trees for Gardens in Southern Africa.* 254 pp. Cape Town: Juta & Co. Illustrated with line drawings, black and white and color photographs.
Watkins, John V. 1975. *Florida Landscape Plants.* Rev. ed. 420 pp. Gainesville: University Presses of Florida. Illustrated with line drawings.

Whitney, Christine M. 1955. *The Bermuda Garden.* 231 pp. Garden Club of Bermuda. Illustrated with line drawings and color photographs.

Wigginton, Brooks E. 1963. *Trees and Shrubs for the Southeast.* 280 pp. Athens: University of Georgia Press. Illustrated with black and white photographs.

Williams, R. O. 1949. *The Useful and Ornamental Plants in Zanzibar and Pemba.* 497 pp. Zanzibar. Illustrated with line drawings, black and white photographs.

Williams, R. O., and R. O. Williams, Jr. 1951. *The Useful and Ornamental Plants in Trinidad and Tobago.* 335 pp. Port-of-Spain: Guardian Commercial Printery.

Woman's Club of Havana. 1951. *Flowering Plants from Cuban Gardens.* 248 pp. Havana: Seoane, Fernandez & Cia. Illustrated with line drawings and watercolor.

Wrigley, John W. 1979. *Australian Native Plants.* 448 pp. Sydney: Collins. A manual for their propagation, cultivation, and use in landscaping. Illustrated with line drawings, black and white and color photographs.

INDEX